Energy Transition in the African Economy Post 2050

Olayinka Ohunakin
Covenant University, Nigeria

A volume in the Practice, Progress, and Proficiency in Sustainability (PPPS) Book Series

IGI Global
PUBLISHER of TIMELY KNOWLEDGE

Published in the United States of America by
IGI Global
Engineering Science Reference (an imprint of IGI Global)
701 E. Chocolate Avenue
Hershey PA, USA 17033
Tel: 717-533-8845
Fax: 717-533-8661
E-mail: cust@igi-global.com
Web site: http://www.igi-global.com

Library of Congress Cataloging-in-Publication Data

Names: Ohunakin, Olayinka, editor.
Title: Energy transition in the African economy post-2050 / Olayinka Ohunakin, editor.
Description: Hershey, PA : Engineering Science Reference, [2022] | Includes bibliographical references and index. | Summary: "This book will provide state and non-state actors with vital information that will deepen their understanding of the Paris Agreement and its impact on economic development in Sub-Saharan Africa, especially in post 2050, when low carbon transition is expected to have been well entrenched in many developed countries"-- Provided by publisher.
Identifiers: LCCN 2022039742 (print) | LCCN 2022039743 (ebook) | ISBN 9781799886389 (hardcover) | ISBN 9781668466186 (library binding) | ISBN 9781799886396 (ebook)
Subjects: LCSH: Sustainable development--Africa, Sub-Saharan. | Power resources--Africa, Sub-Saharan. | Energy policy--Africa, Sub-Saharan. | Climatic changes--Africa, Sub-Saharan. | United Nations Framework Convention on Climate Change (1992 May 9). Protocols, etc. (2015 December 12)
Classification: LCC HC800.Z9.E5 E64 2022 (print) | LCC HC800.Z9.E5 (ebook) | DDC 338.967--dc23/eng/20220830
LC record available at https://lccn.loc.gov/2022039742
LC ebook record available at https://lccn.loc.gov/2022039743

This book is published in the IGI Global book series Practice, Progress, and Proficiency in Sustainability (PPPS) (ISSN: 2330-3271; eISSN: 2330-328X)

British Cataloguing in Publication Data
A Cataloguing in Publication record for this book is available from the British Library.

Practice, Progress, and Proficiency in Sustainability (PPPS) Book Series

Ayman Batisha
International Sustainability Institute, Egypt

ISSN:2330-3271
EISSN:2330-328X

MISSION

In a world where traditional business practices are reconsidered and economic activity is performed in a global context, new areas of economic developments are recognized as the key enablers of wealth and income production. This knowledge of information technologies provides infrastructures, systems, and services towards sustainable development.

The **Practices, Progress, and Proficiency in Sustainability (PPPS) Book Series** focuses on the local and global challenges, business opportunities, and societal needs surrounding international collaboration and sustainable development of technology. This series brings together academics, researchers, entrepreneurs, policy makers and government officers aiming to contribute to the progress and proficiency in sustainability.

COVERAGE

- Innovation Networks
- Technological learning
- Socio-Economic
- ICT and knowledge for development
- E-Development
- Global Business
- Outsourcing
- Intellectual Capital
- Strategic Management of IT
- Green Technology

IGI Global is currently accepting manuscripts for publication within this series. To submit a proposal for a volume in this series, please contact our Acquisition Editors at Acquisitions@igi-global.com or visit: http://www.igi-global.com/publish/.

The Practice, Progress, and Proficiency in Sustainability (PPPS) Book Series (ISSN 2330-3271) is published by IGI Global, 701 E. Chocolate Avenue, Hershey, PA 17033-1240, USA, www.igi-global.com. This series is composed of titles available for purchase individually; each title is edited to be contextually exclusive from any other title within the series. For pricing and ordering information please visit http://www.igi-global.com/book-series/practice-progress-proficiency-sustainability/73810. Postmaster: Send all address changes to above address.

Titles in this Series

For a list of additional titles in this series, please visit: http://www.igi-global.com/book-series/

Handbook of Research on Socio-Economic Sustainability in the Post-Pandemic Era
Jozef Oleński (High School Technology and Economics, Jaroslaw, Poland) Jeffrey Sachs (Columbia University, USA) Masayuki Susai (Nagasaki University, Japan) Yannis Tsekouras (University of Macedonia, Greece) and Arjan Gjonça (London School of Economics and Political Science, UK)
Information Science Reference • © 2023 • 401pp • H/C (ISBN: 9781799897606) • US $270.00

Financing Regions Toward Sustainability in the Midst of Climate Change Risks and Uncertainty
Beata Zofia Filipiak (University of Szczecin, Poland) Dominika Kordela (University of Szczecin, Poland) and Izabela Nawrolska (University of Szczecin, Poland)
Information Science Reference • © 2023 • 364pp • H/C (ISBN: 9781668476208) • US $225.00

Multidisciplinary Approaches to Sustainable Human Development
Vishal Jain (Sharda University, India) and Vinay Kukreja (Chitkara University, India)
Information Science Reference • © 2023 • 382pp • H/C (ISBN: 9781668482230) • US $225.00

5G, Artificial Intelligence, and Next Generation Internet of Things Digital Innovation for Green and Sustainable Economies
Patricia Ordóñez de Pablos (University of Oviedo, Spain) and Xi Zhang (Tianjin University, China)
Engineering Science Reference • © 2023 • 256pp • H/C (ISBN: 9781668486344) • US $280.00

Handbook of Research on Bioeconomy and Economic Ecosystems
Ana Pego (Universidade Nova de Lisboa, Portugal)
Business Science Reference • © 2023 • 502pp • H/C (ISBN: 9781668488799) • US $295.00

IGI Global
PUBLISHER of TIMELY KNOWLEDGE

701 East Chocolate Avenue, Hershey, PA 17033, USA
Tel: 717-533-8845 x100 • Fax: 717-533-8661
E-Mail: cust@igi-global.com • www.igi-global.com

Table of Contents

Detailed Table of Contents

in the Democratic Republic of Congo, Kenya, and Nigeria. The three countries have unique approaches to addressing climate change due to their unique areas of intensive carbon emissions. They represent the East, West, and Southern areas in SSA. The chapter investigates the impact of climate change mitigation activities in line with the Paris Agreement on societies in the chosen countries. It evaluates the core climate change mitigation target per country to identify opportunities and challenges to achieving the mitigation objectives of the Paris Agreement in SSA. The chapter finds that while there are opportunities for SSA countries to pursue the low carbon and climate change mitigation objectives of the Paris Agreement, a danger may lie in following the global climate agenda under the Paris Agreement at the cost of national circumstances.

Power supply in Nigeria has been epileptic due to low capacity for electricity generation, transmission, and distribution. The country's peak dispatch power stands at 5,552.8 MW against estimated demand of about 98,000MW thus stifling socio-economic activities with attendant adverse effect on economic growth. The absence of effective integrated resource planning was identified as a major cause of the low level of power supply in Nigeria as available energy resources have not been optimally harnessed. Using a future scenario approach, this chapter developed plausible scenarios for Nigeria's power supply by the Year 2050 based on a predictive model of seven per cent (moderate growth), ten per cent (high growth) and 13 per cent (optimistic growth) used in analyzing the impact of energy growth on Nigeria's economy. This led to the generation of scenario-based strategies including the optimal diversification of electricity generation mix and restructuring of the power grid system in Nigeria.

This chapter examines the outlook for energy transitions in São Tomé and Príncipe, a small island developing state in Africa. It considers diverse dimensions, including the country's energy profile, national policies, institutions, and emerging challenges and opportunities. Additionally, this chapter discusses the role of international cooperation. The author poses the following research question: what are the challenges faced by São Tomé and Príncipe in the energy transitions? To answer this, the chapter draws on a range of sources, including scientific and grey literature, as well as official national development plans and reports. São Tomé and Príncipe's unique geographical location, political commitment to combating global warming, and membership in key global and regional organizations position it favorably for energy transitions. By underscoring the limitations and difficulties of national transitions in the context of São Tomé and Príncipe, this chapter aims to contribute to a broader understanding of energy transitions in small island developing states.

The Middle East and North Africa region has become one of the most important regions globally due to its energy reserves and geopolitical position. International organizations and researchers have discussed the challenge of sustainable development in the MENA region for many years. In this chapter, by examining the causal relationships between the four components of sustainable development, it is tried to understand better the interaction between sustainable development components in the region. Therefore, by collecting 92 institutional, bio-environmental, economic, and social variables, development indicators were constructed using the principal component analysis method for 2000-2020. This relationship was then tested using a Granger causality model and a dynamic data panel model. The results of this study show well that the economic development achieved at the expense of environmental degradation in this region failed to improve the non-economic components of sustainable development and provided the basis for their decline.

Peer-to-peer (P2P) energy trading using blockchain is presented as a great innovative potential to promote rural electrification. Opportunities and challenges assessment for the implementation of this technology in Sub-Saharan Africa shows that it is only at its embryonic stage in the region. The decreasing cost of stand-alone solar technology and the expansion of investment in mini-grid sector are among the opportunities. However, the considerable restriction of private participation in the mini-grid sector, the difficulty of the regulatory process and licensing requirements, the issues with tariff framework, and the uncertainty of the regulation about the future grid integration are among the main challenges. This chapter proposes a policy and regulation framework for the promotion of P2P energy trading using blockchain in Sub-Saharan Africa.

There is ample evidence in the literature that developing countries would suffer the most from the adverse effects of climate change. Although, respective developing economies have dedicated action plans to mitigate or adapt to these adverse effects, financing for these strategies may be lacking or national governments may not commit financial resources to actualizing these strategies. Using a budget analysis and climate budget tagging framework, the chapter evaluates the financial resources the Nigerian government has committed to its adaptation strategies as stipulated in the 2011 National Adaptation Strategy and Plan of Action on Climate Change (NASPA – CCN). The study found out amongst others that government expenditure on climate change tends to be more of mitigation than adaptation. In addition, adaptation programs targeted at the industry, commerce, telecommunications, and transport sector are most neglected among other sectors highlighted as priority sectors in the NASPA – CCN policy.

In the last few decades, we have seen remarkable growth in the technology sector. But this growth has cost us in the form of global warming and climate change. Information and communication technology (ICT) is the quickest grown area in the technology sector and impacted a lot on every part of our lives and is somehow responsible for global warming and climate change in the form of energy emissions. The increasing level of emission of carbon dioxide (CO2) and other gases from various sources has has great affect. ICT, RES, and strict governance can help to mitigate the problem and contribute to energy efficiency by becoming smart. It can contribute to developing smart homes, smart grids, smart logistics, and other smart devices. Using renewable energy sources (RES), small and medium scale systems, computers and related peripherals, and electric and solar vehicles may contribute to conserving the environment and mitigate climate change and global warming. If we can work honestly, we can secure our future generations by making technologies smarter by integrating RES.

Preface

This book offers a predominant view of the impact of energy transition on the economy of Africa after 2050. Africa as a continent is anticipated to grow in population to approximately 2.5 billion people by 2050. Sub-Saharan Africa is accounted to accommodate the larger percentage of the population. Worthy of note is the fact that in sub-Saharan Africa today, a minority of the entire population has access to electricity (about 600 million without access to electricity), and much lower have access to clean cooking fuels and technologies (about 900 million without access). The progress and some of the gains made in recent years to reduce the energy access deficit via energy transition in Africa, have been hampered by the economic impact of the COVID-19 pandemic. The COVID-19 pandemic has further widened the gap. Africa is also adjudged to be among the most vulnerable parts of the world battling the harsh consequences of climate change despite its low contribution to the emission of CO_2. In addition to all these challenges, Africa is bequeathed to rising insurgencies, insecurity, military coup d'état truncating democratic structures, armed militancy, kidnappings, in most of the regions, etc. The rate of penetration of energy transition to achieving net zero by 2050, might thus have been altered and reduced drastically.

Energy transition through a direct shift to renewable energy (especially in solar, bioenergy, wind, and hydropower), energy efficiency, and other energy transition-related technologies should be viewed as a prospect to improve human existence in Africa rather than just for economic benefits. It is expected to become a major job creation opportunity for Africa by creating many new jobs across sectors and beyond energy. A 20-fold increase by 2050 from today's values in job creation is expected through energy transition, as revealed by IRENA *Could the Energy Transition Benefit Africa's Economies?*

As the world embraces energy transition through greater decarbonization, it is essential that the challenges associated with access to clean, affordable, and inclusive energy plaguing Africa be tackled to achieve a fair and just energy transition (that considers environmental preservation and climate change impact on livelihoods). Africa is endowed with vast renewable energy potential that is making the world appeal to the continent to accelerate the development of its massive renewable energy sources; while this plea is necessary and acceptable, Africa must also be given time to enable a gradual transition starting with the use of its natural gas as a transition fuel while exploring its renewables, just like Germany and other countries in Europe. It is also worth knowing that Africa is a gas continent. Investing in natural gas as a transition fuel will provide Africa with the necessary impetus for rapid growth and development thus taking the continent to the position of genuine competition with the rest of the developed world.

In Chapter 1, Mark M. Akrofi, Nadia I. Rabehi, Donald C. Abonyi, Niyonzima J. Damascene, and Olayinka S. Ohunakin in their work entitled "Energy Diversification in Africa: Status and Implications for the Clean Energy Transition," established and analyzed the status of energy diversification in Africa from 2000 to 2017. Using the Energy Mix Concentration Index method, energy diversification indices were developed for 53 countries and territories for which data were available from the African Union Energy Commission's database. This chapter showed that the electricity generation mix in Africa has moderately diversified between 2000 and 2017. While many countries have introduced renewable energy sources into their generation mix, the share of energy from these sources remains low compared to their fossil counterparts. The chapter concluded that more investment in renewable energy is essential to engender energy diversification, improve energy security, and foster clean energy transitions on the continent.

Adejumoke J. Akinbusoye, Dilinna Lucy Nwobi, Alexander O. Akolo, and Oluwatomiwa P. Phillips in their work in Chapter 2 examined social and legal dimensions of the implementation of climate change mitigation projects and policies in selected sub-Saharan African countries including Nigeria, the Democratic Republic of Congo, and Kenya. The authors observed that while these countries are consistent in their attempts to fulfill the mandates of the international climate agenda, inherent cultural, social, and legal concerns exist that serve as barriers to advancement in line with the goals of the Paris Agreement. Such barriers include poor citizen inclusion in the design of climate change-related laws, policies, and projects and side-lining of indigenous people as well as the violation of their human rights in Kenya, political instability and civil unrest in DRC, and the absence of a renewable energy law and poor implementation of existing laws in Nigeria. On the other hand, the chapter submits that harnessing low-carbon energy opportunities in line with the climate agenda can help to address energy poverty and other forms of poverty in the region. Finally, the chapter concluded that the rich forests and indigenous expertise in sub-Saharan Africa are assets, therefore an inclusive approach can help to improve climate governance by harnessing indigenous expertise in forest management and other climate projects.

In Chapter 3, Olutosin A Ogunleye reported that power supply in Nigeria has been epileptic due to low capacity for electricity generation, transmission, and distribution. The country's peak dispatch power stands at 5,552.8 MW against estimated demand of about 98,000MW thus stifling socio-economic activities with an attendant adverse effect on economic growth. It was identified in the chapter that the absence of effective integrated resource planning was a major cause of the low level of power supply in Nigeria, as the available energy resources have not been optimally harnessed. Using a future scenario approach, the work developed plausible scenarios for Nigeria's power supply by 2050 based on a predictive model of 7 percent (moderate growth), 10 percent (high growth), and 13 percent (optimistic growth) used in analyzing the impact of energy growth on Nigeria's economy. This led to the generation of scenario-based strategies including the optimal diversification of the electricity generation mix and restructuring of the power grid system in Nigeria.

João Simões in his work in Chapter 4 examined the outlook for energy transitions in São Tomé and Príncipe, a small island developing state in Africa. The chapter considers diverse dimensions, including the country's energy profile, national policies, institutions, and emerging challenges and opportunities. Additionally, this chapter discusses the role of international cooperation. The following research questions were posed: What are the challenges faced by São Tomé and Príncipe in the energy transitions? To answer this, the chapter draws on a range of sources, including scientific and grey literature, as well as official national development plans and reports. São Tomé and Príncipe's unique geographical location, political commitment to combating global warming, and membership in key global and regional organizations position it favorable for energy transitions. By underscoring the limitations and difficulties of national transitions in the context of São Tomé and Príncipe, this chapter aims to contribute to a broader understanding of energy transitions in small island developing states.

The Middle East and North Africa region has become one of the most important regions globally due to its energy reserves and geopolitical position. This was expressed by Nima Norouzi in Chapter 5. International organizations and researchers have discussed the challenge of sustainable development in the MENA region for many years. In this chapter, by examining the causal relationships between the four components of sustainable development, the interactions between sustainable development components in the region were better examined. By collecting 92 institutional, bio-environmental, economic, and social variables, development indicators were constructed using the principal component analysis method for 2000-2020. This relationship was then tested using a Granger causality model and a dynamic data panel model. The results of this study show explicitly that the economic development achieved at the expense of environmental degradation in this region failed to improve the non-economic components of sustainable development and provided the basis for their decline.

A peer-to-peer (P2P) energy trading using blockchain is presented in Chapter 6 by Mirana N. Andriarisoa, Erick G. Tambo, David Tsuanyo, and Axel N. Nguedoung, as a great innovative potential to promote rural electrification. Opportunities and challenges associated with the implementation of this technology in sub-Saharan Africa show that the innovative technology is only at its embryonic stage in the region. The decreasing cost of stand-alone solar technology and the expansion of investment in the mini-grid sector are among the opportunities. However, the considerable restriction of private participation in the mini-grid sector, the difficulty of the regulatory process and licensing requirements, the issues with the tariff framework, and the uncertainty of the regulation about future grid integration are among the main challenges. This chapter proposes a policy and regulation framework for the promotion of P2P energy trading using blockchain in Sub-Saharan Africa.

In Chapter 7, Chukwuemeka O. Onyimadu, and Daniel U. Sunday proved that there is ample evidence in the literature that developing countries would suffer the most from the adverse effects of climate change. Although, respective developing economies have dedicated action plans to mitigate or adapt to these adverse effects, financing for these strategies may be lacking or national governments may not commit financial resources to actualize these strategies. Using a Budget Analysis and Climate Budget Tagging Framework, this chapter evaluates the financial resources the Nigerian government has committed to its adaptation strategies as stipulated in the 2011 National Adaptation Strategy and Plan of Action on Climate Change (NASPA – CCN). The study observed amongst others that, government expenditure on climate change tends to be more mitigation than adaptation. In addition, adaptation programs targeted at the industry, commerce, telecommunications, and transport sectors are most neglected among other sectors highlighted as priority sectors in the NASPA–CCN policy.

In the last few decades, growth in the technology sector has been remarkable. Ahmad T. Siddiqui, and Nupur Soni in Chapter 8 explained that this growth has cost a lot in the form of global warming and climate change. Information and Communication Technology (ICT) witnessed the fastest growth area in the technology sector and impacted a lot on every part of our lives and somehow responsible for global warming and climate change in the form of emissions through energy utilization. The increasing level of emissions of carbon-dioxide (CO_2) and other gasses from various sources has been exhilarating. ICT, renewable energy sources (RES), and strict governance can help to mitigate the problem and contribute to energy efficiency by embracing smart approaches. It can contribute to developing smart homes, smart grids, smart logistics, and other smart devices. RES, small and medium scale systems, computers and related peripherals, and electric and solar vehicles may contribute to conserving the environment and mitigate climate change and global warming. With thorough commitment to work, we can secure our future generations by making technologies smarter by integrating RES.

The editor wishes to conclude that this book tends to contribute to a broader discussion on the ways in which energy transition will impact the African economy after net zero emissions by 2050. Energy transition in Africa is still growing and its implementation is defined around seven strategic objectives as revealed in *African Energy Transition Programme*. The academic literature in existence on energy transition in Africa is thus still limited in scope. Worthy of note is the fact that this book has not ended the debate around the ongoing energy transitions and the decarbonization of Africa.

It is also pertinent to say here that the unexpected political unrest and military takeovers of governments in African countries like Niger Republic, Mali, Burkina Faso, and Gabon, and militant insurrections together with armed conflicts in some other regions of Africa will most likely bring about a negative impact on the African economy which is likely to stall the rate of energy transition pre- and post-2050; the extent of this in future may be difficult to determine as at present. All these thus make the scope of this book and depth of analysis to be limited, since energy transition will continue to invite continuous interests, debates, and discussion. The editor selected eight authors across different fields in Africa and another developing country. Every chapter in this work was fully peer-reviewed by reviewers made up of qualified academic and professional experts alongside the editor. The concepts, thoughts, and ideas in the manuscripts belong to the respective authors of the manuscripts and are majorly considered academic information.

Finally, I wish to express my profound gratitude and deep appreciation to all the authors for their time, sacrifice, and knowledge to ensure the success of this project.

Olayinka S. Ohunakin
Covenant University, Nigeria

REFERENCES

Accelerating Africa's Energy Transition. (2023). *Premium Times*. https://www.premiumtimesng.com/promoted/594860-accelerating-africas-energy-transition.html

African Development Bank Group. (2022) *AEC: Accessing clean, affordable energy in Africa is key as the world moves towards energy transition: panelists*. AFDB. https://www.afdb.org/en/news-and-events/2022-aec-accessing-clean-affordable-energy-africa-key-world-moves-towards-energy-transition-panelists-57344

African Energy Transition Programme. (n.d.). African Union. https://au-afrec.org/energy-transition-programme

Ferroukhi, R. (2022). *Could the Energy Transition Benefit Africa's Economies?* IRENA. https://www.irena.org/News/expertinsights/2022/Nov/Could-the-Energy-Transition-Benefit-Africas-Economies#:~:text=The%20energy%20transition%20has%20the%20potential%20to%20boost%20employment%20in,by%202050%20under%201.5%2DS

Ighobor, K. (2022). *A just transition to renewable energy in Africa.* UN. https://www.un.org/africarenewal/magazine/november-2022/just-transition-renewable-energy-%C2%A0africa

OpecFund. (2021). *The path to Africa's energy transition.* Opec Fund. https://opecfund.org/news/the-path-to-africa-s-energy-transition

Chapter 1
Energy Diversification in Africa:
Status and Implications for the Clean Energy Transition

Mark M. Akrofi
https://orcid.org/0000-0003-2516-2172
United Nations University, Japan

Ilhem N. Rabehi
Zian Achoure University, Algeria

Donald C. Abonyi
Technical University of Munich, Germany

Niyonzima J. Damascene
Hydro Operation Great Lakes, Kigali, Rwanda

Olayinka S. Ohunakin
Covenant University, Nigeria

ABSTRACT

In this chapter, the authors establish and analyze the status of energy diversification in Africa from the year 2000 to 2017. Using the energy mix concentration index method, energy diversification indices were developed for 53 countries and territories for which data were available from the African Union Energy Commission's database. The researchers find that the electricity generation mix in Africa has moderately diversified between 2000 and 2017. While many countries have introduced renewable energy sources

DOI: 10.4018/978-1-7998-8638-9.ch001

into their generation mix, the share of energy from these sources remains low compared to their fossil counterparts. More investment in renewable energy is essential to engender energy diversification, improve energy security, and foster clean energy transitions on the continent.

1. INTRODUCTION

Following the launch of the Sustainable Development Goals (SDGs) and the Paris Agreement in 2015, there has been increased international calls for countries to diversify their energy mix, essentially by phasing out fossil fuels and replacing them with renewable energy sources. This transition requires balancing the need for renewables with economic growth and energy security, given that most economies around the globe are heavily driven by fossil fuels. Africa has experienced immense economic growth over the past two decades, and it is expected to witness more growth over the coming ones. For a sustainable future, significant efforts are required to ensure that the continent's economic growth is driven by clean energy while advancing energy access and ensuring energy security. Energy diversification, especially through the development of renewable energy resources is an essential strategy and a precursor for decarbonization, clean energy transition, and energy security (Stirling, 2008). Countries and regions have set ambitious goals and targets to achieve this transition, with the European Union, for example aiming to achieve 32% of renewable energy in its energy mix by 2030 (De et al., 2019). Other advanced economies have set similar targets and are committing enormous resources towards this goal. Significant progress on improving energy access and transition has also been made in Africa; yet, its growing population and economic progress has sent energy demand soaring (IRENA, 2015).

Thus, there is an urgent need for a rapid increase in energy supply, to which all forms of energy must be included in production in the decades ahead. Africa is very rich with renewable energy sources. However, the continent still faces a reliance on gas, oil and traditional biomass. The African Union's Agenda 2063 aims to achieve high standards of living, quality of life and well-being for all citizens in a peaceful and stable Africa as well as to foster economic expansion accompanied by the full achievement of the Sustainable Development Goals (SDGs) by 2030 (IEA, 2019). It also aims to diversify energy resources, achieve full access to electricity and clean cooking and to reduce stigmatically premature deaths related to pollution (IEA, 2019). Electricity generation in Africa increased to 870 TWh in 2018 from 670 TWh in 2010 (IEA, 2019). Natural gas and coal (the latter largely in South Africa) accounted for 40% and 30% of generation output in 2018 respectively, while Hydropower accounted for a further 16% and oil for 9% (IEA, 2019).

Renewable power capacity increased from 28 GW in 2010 to almost 50 GW in 2018 (IEA, 2019). Hydropower is the largest source of renewable power by far and its capacity increased from 26 GW in 2010 to 35 GW in 2018, although its share in the overall generation mix has remained relatively constant at around 15% (IEA, 2019). Other renewable sources have started to develop but, for the moment, their share in generation and capacity is low. However, large regional variations occur in terms of the composition of the energy mix. There are also variations in the distribution of energy resources in the continent. Hence, various context-specific factors could influence the diversification and transition pathways of the different countries in the continent. With the urgency of the climate crisis, the need for diversification, particularly through the development of renewable energy is well-understood and currently features prominently in both countries, regional and global-level policies. However, studies on energy diversification in Africa are still limited and diversity has not been studied from a long-term comparative perspective, especially using concentration indicators that enable cross-country comparisons, in Africa (Akrofi, 2021).

Hence, the aim of this chapter is to establish and analyze the status of energy diversification for all African countries and territories for which data is available. The scope of the study covers the period from the year 2000 to 2017 based on available data from the African Union Energy Commission (AFREC). We compute energy diversification indices for the year 2000 and 2017, and provide a comprehensive view of how the energy mix was diversified over this 18-year period. A regional-based analysis was also done to provide more insights into the diversification status of each of the five sub-regions in Africa (Western, Northern, Southern, Eastern and Middle Africa). The rest of the chapter is organized as follows. In section 2, the method adopted for the study is explained, in section 3 the results and discussions are presented while the implications and conclusions are presented in section 4.

2. MATERIALS AND METHODS

The study relies on secondary data obtained from the AFREC data portal[1]. AFREC is an agency dedicated to keep an updated energy database for all African countries. For this study, data on electricity generation regarding the amount of electricity generated (in Megawatt hours—MWh) from each energy source in the energy mix of each country was collected for all African countries. Variety and balance (Stirling, 2008) in the energy mix were considered in computing the energy diversification indices. Variety refers to the number of energy sources (coal, natural gas, solar, etc.) in the energy mix, while balance refers to the amount of electricity generated from each energy source (Rubio-Varas & Muñoz-Delgado, 2019a; Stirling, 2008). To compute the diversification indices, the Energy Mix Concentration Index Method (Rubio-Varas & Muñoz-Delgado, 2019b) was adopted. The EMCI method is a modified version of the Herfindahl–Hirschman Index (HHI) which has traditionally been applied to measure market competitiveness and concentration in an industry regarding the size of each firm in relation to the industry (Akrofi, 2021).

The EMCI is the summation of the squares of the proportion of each energy source in the energy mix in any given period (Rubio-Varas & Muñoz-Delgado, 2019b) and it is expressed as:

$$EMCI_t = \sum_i^t P_i^2 \tag{1}$$

Where P_i is the share of the energy source i in the energy mix. Thus, i represents the different energy sources in the energy mix (Natural Gas, Coal, Hydro, Geothermal, etc.). Balance (the share of each energy source- P_i) is expressed as a fraction in this study (that is, the amount of energy generated from a particular source divided by the amount of electricity generated from all sources in a particular year). Fractions were maintained instead of percentages for the purpose of simplicity in computing the EMCIs since the EMCI is a summation of the squares of the share of electricity from each energy source in the energy mix. Hence, the square of each share was taken for each energy source and summed up to arrive at the EMCI. The value of the EMCI ranges between 0 and 1, where 0 indicates total diversity and 1 indicates no diversity (Rubio-Varas & Muñoz-Delgado, 2019a).

3. RESULTS AND DISCUSSION

3.1. Status of Diversification in the Energy Generation Mix of African Countries

The results unearth a gradual shift from a highly concentrated energy generation mix in the 2000s to a more diversified energy mix as at 2017. Using cut points of 0-0.40, 0.41-6.0 and 6.1+ for high, moderate, and low, only four countries (Mauritius, Kenya, Sierra Leone and Guinea) qualified as highly diversified in 2000 with 14 being moderately diversified and the majority (36) being least diversified. However, by 2017, these proportions have changed to 7, 22, and 25 for high, moderate and least diversified countries, indicating a decline from 66.7% in the year 2000 to 46.3% in the year 2017 for least diversified countries. Burkina Faso emerged as the most diversified country as at 2017 and other countries such as Cote d'Ivoire, Uganda and Morocco were among the new countries to have graduated into this highly diversified category. Figure 1 shows the status of diversification in 2000 and 2017.

Characteristically, all highly diversified countries as at 2017 have a higher variety of energy sources in their energy generation mix, with countries in this category having at least five different energy sources. Morocco and Gabon had the highest variety with seven different energy sources each. On the other hand, most countries in the least diversified category are characterised by a lower variety (few energy sources). Countries such as The Gambia (2 energy sources), Eritrea (2 energy sources), and South Sudan (3 energy sources)

Figure 1. Diversity in the energy generation mix of African countries in year 2000 and 2017[2]
Source: Authors' construct

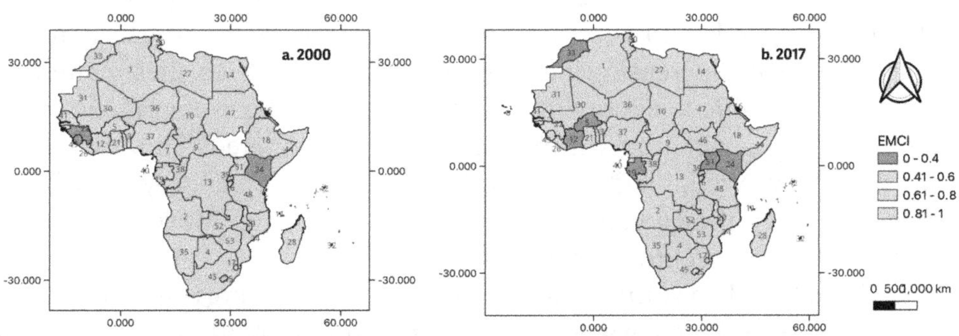

OBJECTID	NAME
1	Algeria
2	Angola
3	Benin
4	Botswana
5	Burkina Faso
6	Burundi
7	Cameroon
8	Cape Verde
9	Central African Republic
10	Chad
11	Comoros
12	Côte d'Ivoire
13	Democratic Republic of the Congo
14	Egypt
15	Equatorial Guinea
16	Eritrea
17	Eswatini
18	Ethiopia
19	Gabon
20	Gambia
21	Ghana
22	Guinea
23	Guinea-Bissau
24	Kenya
25	Lesotho
26	Liberia
27	Libya
28	Madagascar
29	Malawi
30	Mali
31	Mauritania
32	Mauritius
33	Morocco
34	Mozambique
35	Namibia
36	Niger
37	Nigeria
38	Republic of Congo
39	Rwanda
40	Sao Tome and Principe
41	Senegal
42	Seychelles
43	Sierra Leone
44	Somalia
45	South Africa
46	South Sudan
47	Sudan
48	Tanzania
49	Togo
50	Tunisia
51	Uganda
52	Zambia
53	Zimbabwe

amongst others have fewer than four energy sources in their energy generation mix. On the first look, this result suggest that countries with a higher variety (number of energy sources in the energy mix) are more diversified than countries with lower variety, and will affirm Stirling's (2008) hypothesis that all things being equal, the higher the variety, the greater diversity.

However, a few countries prove to be an exception to this assertion. Least diversified countries such as DRC (6 energy sources), Mozambique (5 energy sources), Namibia (5 energy sources), and Tunisia (6 energy sources) amongst others have a higher variety of energy resources in their energy mix. This exception is attributed to the share of electricity generated from each energy source. The most diversified countries have a wide spread in terms of the share of energy generated from each source in the energy mix while the share of electricity generated from each source is highly skewed towards a single energy source in the energy mix of the least diversified countries as portrayed in Figure 2.

The findings above suggest that the higher the balance (the share of energy generated from each source), the greater the diversity as postulated by Stirling (2008). The findings affirm those made by Akrofi (2021a) who also found that balance significantly influenced the level of diversification in the energy mix of Africa's top ten economies. It can be deduced from Figure 2, that the least diversified countries are also prone to energy insecurity than the highly diversified ones based on the balance in the energy mix. Least diversified countries such as Lesotho, DRC, Central African Republic (CAR), and Zambia depend almost entirely on Hydro for their electricity. In a time of climate change and increasing incidents of droughts in Africa, these countries are more susceptible to energy insecurity because in the event of a drought, their electricity generation will be significantly hampered. This viewpoint is supported by Azzuni and Breyer's (2020) findings on energy security in which least diversified countries such as those mentioned above also feature in the least energy secure countries in the world.

Fossil fuels still constitute a large proportion of the energy mix of African countries. However, renewable energies are gaining traction, especially in sub-Sahara Africa. Hydro power in particular remains the largest source of renewable power by far and its capacity increased from 26 GW in 2010 to 35 GW in 2018, although its share in the overall generation mix has remained relatively constant at around 15% (IEA, 2019). North and South Africa in contrast are hugely reliant natural gas (North Africa), coal and to a modest

Figure 2. Shares of energy sources in the electricity generation mix of the most diversified (a) and least diversified (b) countries
Source: Authors' construct

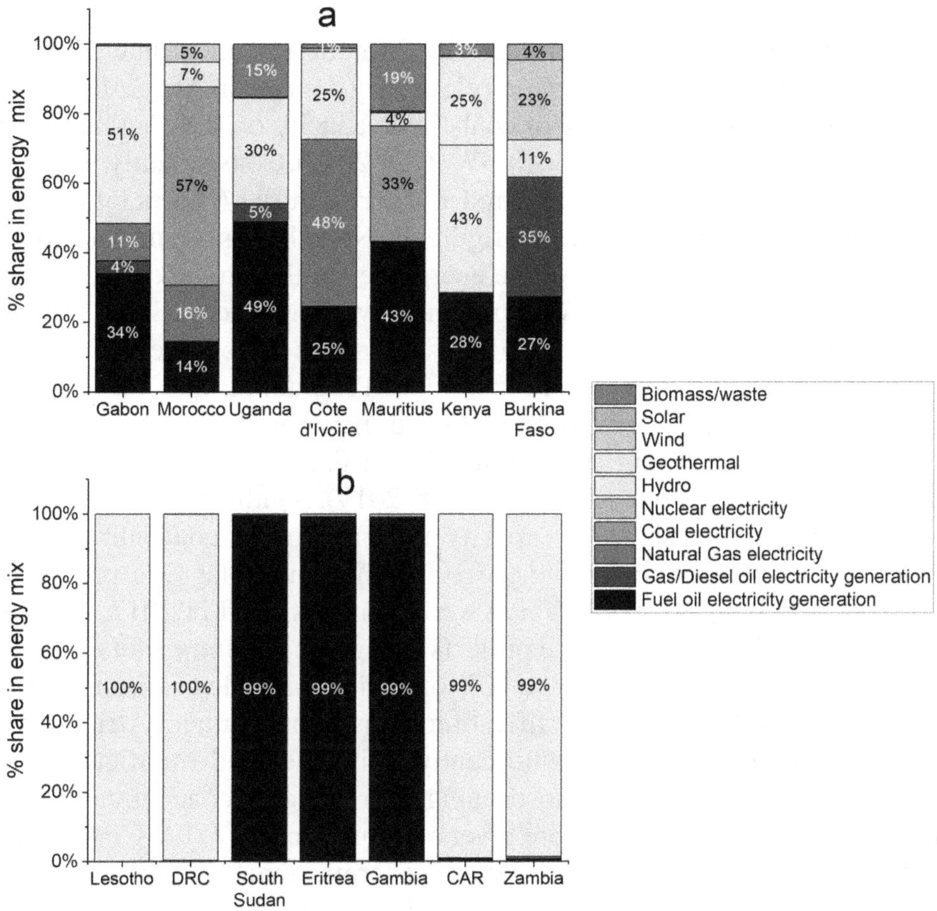

extent on nuclear power (South Africa) while in the remainder of sub-Saharan Africa, hydropower provides over half of generation output, with oil and gas accounting for most of the remaining balance. A regional level analysis of energy diversification in the five regions of Africa is given in the ensuing sections.

3.2. Diversity in the Energy Mix by Sub-Region

3.2.1. Southern Africa

South Africa is hugely reliant on coal and to a modest extent on nuclear power, and it accounts for about 30% of total electricity demand in Africa. It accounts for 85% of the almost 50 GW of coal-fired capacity on the continent while oil-fired capacity totals just over 40 GW (IEA, 2019). Consequently, it remains relatively less diversified as compared to other countries (except for Lesotho) in the southern Africa region (see figure 3). Nonetheless, significant efforts are being made to boost renewable energy generation in the country. South Africa accounts for around 2 GW of wind power in Africa (IEA, 2019). The growth of wind power in the country is in part due to its Renewable Energy Independent Power Producer Procurement Programme that was launched in 2011 and has delivered close to 3 GW of new capacity. Notable projects include the Loeiresfotein and Khobab Wind Farms (140 megawatts each) which were commissioned in 2017 (Malik, 2019). South Africa also has close to 2 GW of installed solar PV capacity and a number of concentrating solar power (CSP) projects including the 100 MW Xina Solar One project and the 100 MW Ilanga-1 plant, which were commissioned in 2017 and 2018, respectively (IEA, 2018). These projects brought the country's total installed CSP capacity to 0.4 GW, close to 40% of Africa's installed capacity of CSP. Figure 3 shows the energy diversification status in the Southern Africa region.

Botswana has improved significantly in its energy diversification over the 18-year period studied, transitioning from a highly concentrated energy mix (EMCI =1) in 2000 to more diversified mix in 2017 (EMCI= 0.48). In 2000, the country depended on only one energy source (fuel oil). However, by 2017, new energy sources such as gas/diesel oil, coal, solar and biomass were added in the energy mix, hence, the country had five energy sources in 2017. Lesotho remains least diversified, with a heavy reliance on hydropower. Even though solar energy has been added in the energy mix in 2017, making it the only different energy source in the energy mix other than hydro, energy generated from this source constitutes a minute proportion in comparison to hydro which provides about 99% of the total electricity. The heavy reliance on hydro puts the country at risk of energy insecurity, given that droughts are becoming more frequent in the southern Africa region (Mahlalela et al., 2020). Other countries like Namibia and Eswatini remain relatively more secure as they depend on different energy sources such as Fuel oil, gas/diesel and coal and solar, even though solar provides a larger proportion of their electricity.

3.2.2. Northern Africa

The Northern Africa region is heavily reliant on oil and gas for its electricity needs. The region is home to almost 85 GW of Africa's 100 GW of gas-fired power plants, with natural gas contributing more than three-quarters of power

Figure 3. Status of energy diversification in Southern Africa in year 2000 and 2017
Source: Authors' construct

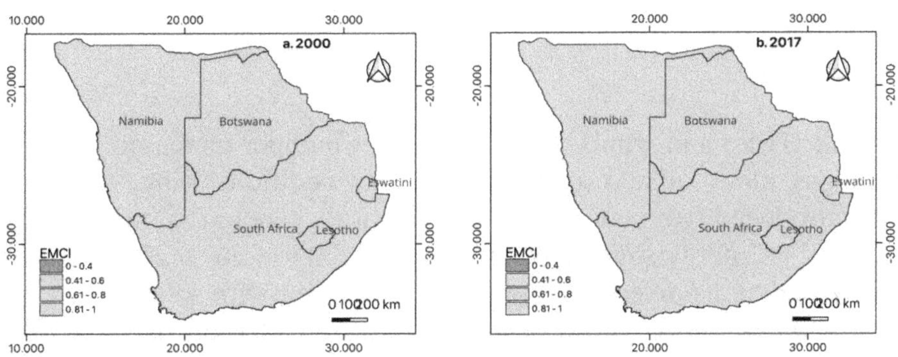

generation in this region in 2018 (IEA, 2019). Nonetheless, countries in this region are making significant strides with regard to renewable energy generation. The region accounts for around 2.6 GW of wind power in Africa while countries such as Tunisia and Morocco are employing the independent power producers (IPPs) model to increase their wind power generation (Malik, 2019). Morocco, Libya and Egypt were the most diversified countries in the region as at 2017 with Morocco remaining a traditionally (considering 2000 as the base year) more diversified country while Libya and Egypt witnessed significant improvements from EMCI's of 0.66 and 0.67, respectively in 2000 to 0.50 and 0.56, respectively in 2017. Figure 4 shows the status of energy diversification in the region.

With significant efforts and investments in renewables, the diversity in Egypt's energy mix could further improve. In 2019, the country's renewable energy generation capacity increased by 1.6GW when the Benban Solar project, one of the largest utility-scale solar PV project on the continent to date, was commissioned. In 2000, the country depended on only three energy sources

Figure 4. Status of energy diversification in Northern Africa in year 2000 and 2017
Source: Authors' construct

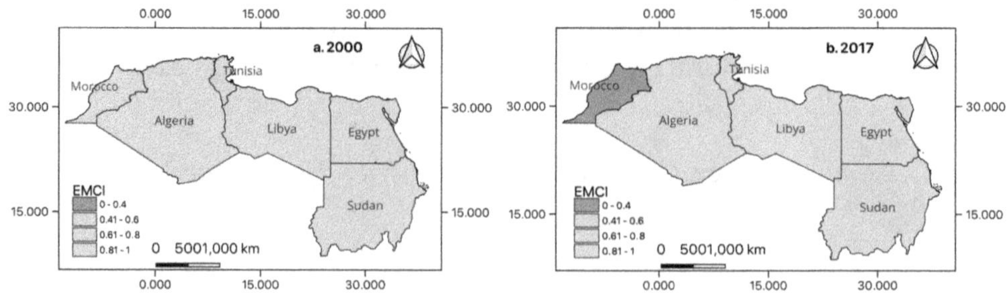

(Fuel oil, Hydro and wind) but by 2017, this number increased to six with natural gas, diesel oil and solar being the new additions. Libya on the other hand, introduced two new energy sources in its generation mix as at 2017 increasing its energy sources from two in 20000 to four in 2017, hence the improvement in its diversification index. However, with solar energy as the only renewable source, Libya's energy mix remains largely fossil fuel-based which is not much of a surprise given that the country is endowed with huge oil resources.

Electricity generation in Algeria, another oil-rich country is also dominated by fossil fuels even though the country is making significant strides in renewable energy generation. As apparent in Figure 4, Algeria remains the least diversified country in the Northern Africa region with no visible changes between the year 2000 and 2017. The country, however, has five energy sources as at 2017, adding two renewable energy sources (wind and solar) to the three sources (Fuel oil, Gas/diesel oil and Natural gas) that existed in 2000. More than 95% of the electricity generated comes from these three sources, which explains the level of concentration in the energy mix.

As noted, the share of energy generated from each source in the energy mix greatly influences the level of diversity other than the number of sources in the energy mix (Akrofi, 2021a). The country's slow pace of diversification, especially with the development of renewables has been attributed to its heavy reliance on the oil and gas sector for economic growth and government subsidies for fossil fuel products (Haddoum et al., 2018). Also, despite having six different energy sources, about 95% of Tunisia's electricity comes from three sources (fuel oil, gas/diesel oil, and natural gas), hence the low level of diversification observed in its electricity generation mix.

3.2.3. Eastern Africa

The East Africa region as a whole witnessed a significant improvement in energy diversification over the 18-year period with countries such as Uganda (EMCI= 0.99), Tanzania (EMCI= 0.80), Rwanda (EMCI= 0.95), and Djibouti (EMCI= 1.00) showing marked changes in their diversification index from a state of high concentration in 2000 to a high level of diversification in 2017 (EMCIs of 0.36, 0.41, 0.44, and 0.51, respectively). Kenya, Madagascar, Mauritius and Sudan which as at the year 2000 were diversified, have continued to improve on their diversification with a change in EMCI's from 0.39, 0.567, 0.35 and 0.50, respectively in 2000 to 0.33, 0.46, 0.34, and 0.48, respectively in 2017. The status of energy diversification in eastern Africa is shown in Figure 5.

The East Africa region is home to some of the continent's massive renewable energy resources. Geothermal resources are generally concentrated in this region where the geothermal energy potential is equivalent to more than 15 GW (Omenda & Teklemariam, 2010). With excellent geothermal resources, Kenya has installed more than 600 MW of capacity while plans are underway to develop an additional 1000 MW from three geothermal projects

Figure 5. Status of energy diversification in Eastern Africa in year 2000 and 2017[3]
Source: Authors' construct.

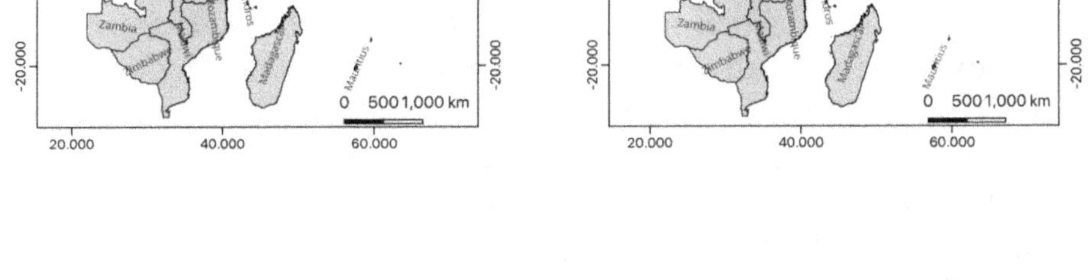

(Geothermal Development Company, 2019). Other countries in East Africa, including Ethiopia, Eritrea, Djibouti, Tanzania and Uganda are also looking to tap their geothermal resources.

With a heavy reliance on fuel oil for their electricity, Eritrea and South Sudan are the least diversified countries in East Africa, followed by Ethiopia which relies heavily on hydro. Even though Ethiopia has five different energy sources in its electricity generation mix in 2017 (having added wind and solar), about 95% of its electricity comes from hydro implying that balance (the share of electricity per source) remains highly skewed towards just one energy source. Hence, the low level of diversification witnessed.

3.2.4. Middle Africa

The middle Africa region also witnessed some improvements in energy diversification from 2000 to 2017. However, while most of the countries have made some improvements, DRC and Central African Republic (CAR) remain the two least diversified countries in this region. Both countries rely heavily on hydro with around 99% of their electricity coming from this source in 2017. DRC has, however, increased its energy sources from two in 2000 to six in 2017. The new energy sources added were Gas/diesel oil, natural gas, solar and biomass. But only a tiny fraction (less than 1%) of the total electricity generated in the country comes from these sources as of 2017. CAR on the other hand introduced only one energy source (Gas/diesel oil) to its original two sources (fuel oil and hydro) in 2000 with over 99% of its electricity coming from the latter. The DRC's dependence on hydro is not surprising given the presence of the Inga Falls which potentially makes it home to the world's largest power station if the seven proposed Inga dams are built. The Inga dams have been projected to produce between 40 to 70GW of electricity if built as planned. The status of energy diversification in the region is shown in Figure 6.

One of the countries that achieved the most significant transformation in its energy diversification is Chad whose EMCI improved from 1.00 in 2000 to 0.53 in 2017. However, the country is dependent on only two energy sources having added Fuel oil as a second source of electricity by 2017. The improvement in its diversification is explained by the fact that there is a relatively more balanced share of electricity generated from each source rather than a skewed balance as seen in the previous countries such as DRC. About 60% of its electricity came from gas/diesel oil while around 40% came from fuel oil in 2017. The country has, however, initiated efforts to generate electricity from renewable sources such as solar and wind. Data from the International Renewable Energy Agency (IRENA, 2020a) shows that Chad has a total installed capacity of 0.190MW for solar and 1.100MW for wind in 2020. Its ministry of finance and energy have also signed an MoU with an

Argentine conglomerate (Alcaal Group) to construct a 200MW solar plant near the capital, N'Djamena (Zyl, 2020). Equatorial Guinea experienced a slight improvement in its diversity over the 18-year period with its EMCI changing from 0.78 in 2000 to 0.56 in 2017. Other countries in the region such as Angola, Gabon, Republic of Congo and Cameroon remain relatively more diversified.

3.2.5. Western Africa

Overall, the Western Africa region witnessed only a slight improvement in energy diversification as compared to the other regions. Some countries such as Burkina Faso (EMCI= 0.26), Cote d'Ivoire (EMCI= 0.36), Togo (EMCI=

Figure 6. Status of energy diversification in Middle Africa in year 2000 and 2017
Source: Authors' construct

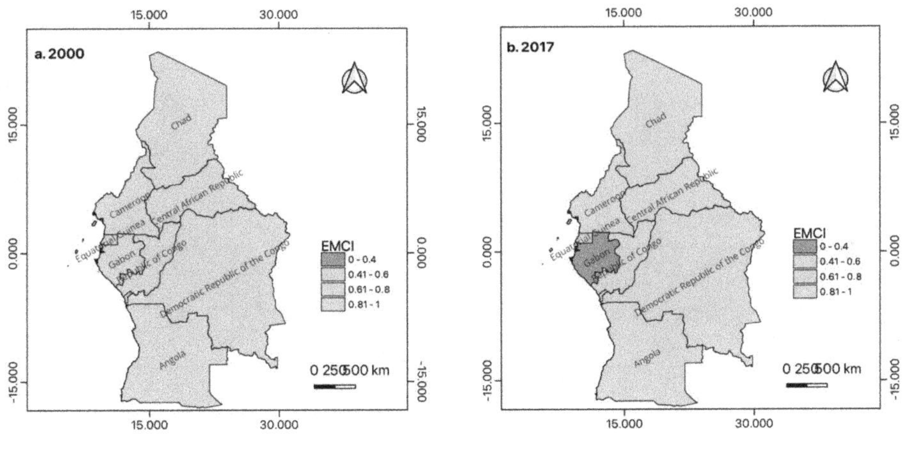

0.44) and Ghana (EMCI= 0.49) have as at 2017 made significant improvements in their diversification, having moved from EMCIs of 0.86, 0.53, 0.51, and 0.84, respectively in the year 2000. On the other hand, countries such as Mali have rather moved towards concentration in their energy mix with a change in the EMCI from 0.50 in 2000 to 0.86 in 2017. Nigeria's EMCI also increased from 0.53 in 2000 to 0.64 in 2017 while no major changes can be observed in countries such as Mauritania and Niger. In the case of Mali, the country depended on three energy sources in 2000 with a relatively more balanced share of electricity from each source. However, despite adding two more energy sources, the balance (share of electricity from each source) became heavily skewed towards hydro which produced around 93% of the total electricity generated, hence the shift from a more diversified mix in 2000 to a high level of concentration in 2017. This case also proves the fact that balance in the energy mix affects the state of diversity other than the variety of energy sources in the energy mix. Figure 7 shows the status of energy diversification in the region.

In the case of Nigeria which relies more on Natural gas, the proportion of electricity from natural gas increased from around 62% in 2000 to about 76% in 2017. This increment explains the reduction in diversification experienced in its energy mix since the share of energy became more skewed towards natural gas. Nonetheless, the country has set some ambitious renewable

Figure 7. Status of energy diversification in Western Africa in year 2000 and 2017
Source: Authors' construct

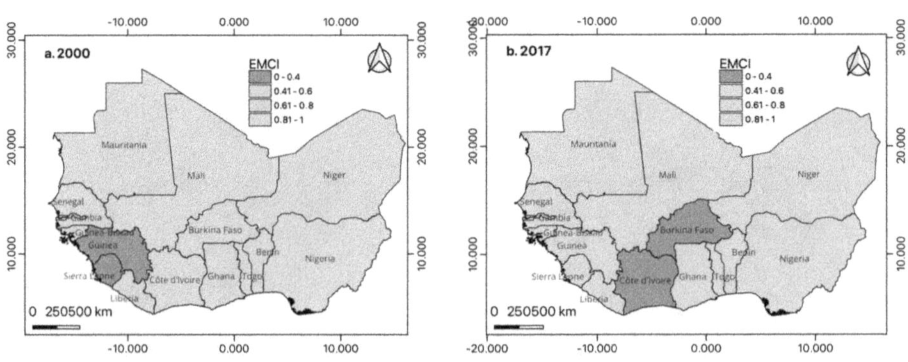

energy targets in recent times which could see its energy mix become more diversified in the coming years. As part of its COVID-19 recovery efforts, the government has launched a new solar strategy which aims to provide around five million solar home systems and mini-grids by 2023 (Akrofi, 2021; Ogunbiyi, 2020). Countries such as Benin, The Gambia, Senegal, and Liberia need to increase the balance in their energy mix and introduce new energy sources, especially renewable energies in order to improve upon their diversity and also boost their energy security. Regional efforts such as the West

Africa Power Pool, the West African Clean Energy Corridor and the ECOWAS Center for Renewable Energy and Energy Efficiency could significantly boost renewable energy development in the region and consequently improve the level of diversification as well as energy security in the region. For instance, ECOWAS has set out an ECOWAS Renewable Energy Policy (EREP) which seeks to increase the share of non-hydro renewable energy in the region's energy mix 48% by 2030 (IRENA, 2020b).

4. CONCLUSION AND RECOMMENDATIONS

4.1. Implications for energy transitions in Africa

Many countries in Africa have introduced at least one renewable energy sources in their energy mix since 2000, with the most common sources being solar, wind, and biomass. Currently, there are significant efforts to boost renewable energy development in the continent. Despite the steady progress being made, the current energy mix is still dominated by fossil fuels. This calls for more action to facilitate the development and utilization of renewable energies, especially given the urgency of climate change, whose effects are already manifesting in the forms of droughts, and other extreme weather events on the continent.

Even though many countries have renewable energy sources in their energy mix, the share of electricity generated from these sources is low as compared to their fossil counterparts. Many countries are also heavily dependent on hydroelectricity. Hence, it is essential for more investments to be made in increasing the share of energy generated, particularly from solar and wind energy. Countries relying on hydro need to diversify their energy mix by introducing other renewables such as wind, solar or geothermal (especially in Eastern Africa) in order to become more energy secure given the likelihood of droughts which could significantly hamper hydro power generation.

4.2. Conclusion

Diversifying the electricity generation mix provides numerous benefits such as improving energy security, and environmental sustainability when fossil fuels are replaced with renewable energies in the energy mix. In Africa, many countries still grapple with improving electricity access despite a vast renewable energy resource potential. Current energy systems are also dominated by fossil fuels. In this chapter, we examined the trajectory of energy diversification on the continent using the year 2000 and 2017 as reference points based on available data. Our study shows that overall, the electricity generation mix on the continent has moderately diversified between 2000 and 2017. As at 2017, seven countries were highly diversified, while 22, and 25 countries were moderately and least diversified, respectively.

The number of least diversified countries declined from 66.7% in 2000 to 46.3% in 2017. It was observed that the share of electricity generated from each energy source has a greater influence on the status of energy diversification other than the number of energy sources in the energy generation mix. It is, thus, paramount for governments to prioritize increasing the share of electricity generated from each energy source, especially renewable energy sources which are abundant in the continent. To conclude, we will like to highlight that while this study provides some insights into energy diversification in Africa, the data used from the AFREC portal is only available from 2000 to 2017 as at the time of conducting this study, hence, our analysis do not extend beyond this period. Also, slight difference maybe observed in data obtained from other databases. Hence, we reiterated that our analysis is solely based on the AFREC data which can be accessed on the AFREC data portal[4].

REFERENCES

Akrofi, M. M. (2021a). An analysis of energy diversification and transition trends in Africa. *International Journal of Energy and Water Resources*, *5*(1), 1–12. doi:10.100742108-020-00101-5

Akrofi, M. M. (2021b). Balancing Post-COVID-19 Economic Growth with Renewable Energy Development in Oil-producing African Economies: Nigeria in Perspective. *Journal of International Affairs*. https://jia.sipa.columbia.edu/online-articles/balancing-post-covid-19-economic-growth-renewable-energy-development-oil-producing

Azzuni, A., & Breyer, C. (2020). Global energy security index and its application on national level. *Energies*, *13*(10), 2502. doi:10.3390/en13102502

De, M., Mata, E., Scholten, D., & Smith, K. (2019). The multi-speed energy transition in Europe : Opportunities and challenges for EU energy security. *Energy Strategy Reviews, 26*(August). doi:10.1016/j.esr.2019.100415

Geothermal Development Company. (2019). *Geothermal Development Company*. GDC. https://www.gdc.co.ke

Haddoum, S., Bennour, H., & Ahmed Zaïd, T. (2018). Algerian Energy Policy: Perspectives, Barriers, and Missed Opportunities. *Global Challenges (Hoboken, NJ)*, *2*(8), 1700134. doi:10.1002/gch2.201700134 PMID:31565341

IEA. (2018). *Market Report Series: Renewables 2018*. IEA. https://www.iea.org/reports/renewables-2018

IEA. (2019). *Africa Energy Outlook 2019 Africa Energy Outlook 2019*. IEA.

IRENA. (2015). [*A Roadmap for Renewable Energy Future*.]. IRENA.

IRENA. (2020a). *Insights on Renewables: Installed Capacity Trends*. IRENA. https://www.irena.org/

IRENA. (2020b). *West Africa Clean Energy Corridor*. IRENA. https://www.irena.org/cleanenergycorridors/West-Africa-Clean-Energy-Corridor

Mahlalela, P. T., Blamey, R. C., Hart, N. C. G., & Reason, C. J. C. (2020). Drought in the Eastern Cape region of South Africa and trends in rainfall characteristics. *Climate Dynamics*, *55*(9–10), 2743–2759. doi:10.100700382-020-05413-0 PMID:32836893

Malik, S. (2019). *Africa's Wind Project Pipeline Grows to 18GW.* Greentech Media. https://www.greentechmedia.com/articles/read/africa-18-gw-wind-project-pipeline

Ogunbiyi, D. (2020, July). *Nigeria is using the pandemic to build a better energy future*. World Economic Forum. https://www.weforum.org/agenda/2020/09/nigeria-using-pandemic-build-sustainable-energy-future/

Omenda, P., & Teklemariam, M. (2010). *Overview Of Geothermal Resource Utilization In The East African Rift System*. RISE. https://rise.esmap.org/data/files/library/kenya/Renewable Energy/RE 9.2 Omenda Teklemariam 2010 Overview of Geothermal Resource Utilization in the East African Rift System.pdf

Rubio-Varas, M., & Muñoz-Delgado, B. (2019a). Long-term diversification paths and energy transitions in Europe. *Ecological Economics*, *163*(June), 158–168. doi:10.1016/j.ecolecon.2019.04.025

Rubio-Varas, M., & Muñoz-Delgado, B. (2019b). The Energy Mix Concentration Index (EMCI): Methodological considerations for implementation. *MethodsX*, *6*(May), 1228–1237. doi:10.1016/j.mex.2019.05.023 PMID:31193910

Stirling, A. (2008). Diversity and sustainable energy transitions: Multicriteria diversity analysis of electricity portfolios. In M. Bazilian & F. Roques (Eds.), *Analytical Methods for Energy Diversity and Security* (pp. 1–29). Elsevier Ltd., doi:10.1016/B978-0-08-056887-4.00001-9

Zyl, N. P. (2020, August 13). *Chad solar power projects earmarked to support N'Djamena surrounds in Chad*. ESI-Africa. https://www.esi-africa.com/industry-sectors/generation/solar/solar-projects-earmarked-to-support-ndjamena-surrounds-in-chad/

ENDNOTES

1 https://au-afrec.org/En/administration/index.php

2 Data were unavailable for South Sudan in the year 2000

3 Data were unavailable for South Sudan in the year 2000

4 https://au-afrec.org/En/administration/index.php

Chapter 2
Socio-Legal Opportunities and Barriers to Achieving the Objectives of the Paris Agreement in Selected Sub-Saharan African Countries

Adejumoke Juliana Akinbusoye
University of Lagos, Nigeria

Alexander Olatunde Akolo
University of Ibadan, Nigeria

Dilinna Lucy Nwobi
University of Ibadan, Nigeria

Oluwatomiwa Priscilla Phillips
University of Ibadan, Nigeria

ABSTRACT

The 2015 Paris Agreement promises to strengthen the global response to climate change in the context of sustainable development. This chapter examines the socio-legal opportunities and barriers to achieving the objectives of the Paris Agreement in the Democratic Republic of Congo, Kenya, and Nigeria. The three countries have unique approaches to addressing climate change due to their unique areas of intensive carbon emissions. They represent the East, West, and Southern areas in SSA. The chapter investigates the impact of climate change mitigation activities in line with the Paris Agreement on societies in the chosen countries. It evaluates the core climate change mitigation target per country to identify opportunities and challenges to achieving the mitigation objectives of the Paris Agreement in SSA. The chapter finds that while there are opportunities for SSA countries to pursue the low carbon and climate change mitigation objectives of the Paris Agreement, a danger may lie in following the global climate agenda under the Paris Agreement at the cost of national circumstances.

DOI: 10.4018/978-1-7998-8638-9.ch002

INTRODUCTION

The global climate agenda involves a broad range of country participation towards limiting global warming to below 2°C, known as the goal to net zero. The United Nations defines net zero as reducing greenhouse gas emissions to nearly 0%, while excess emissions are absorbed from the atmosphere by carbon sinks, including the oceans and forests. (Article 4, (1b) UNFCCC).

The 2015 Paris Agreement promises to strengthen international climate change mitigation and adaptation efforts. While there are general opportunities for promoting the goals of the international climate change agenda in Sub-Saharan African countries. The current energy crisis is a major opportunity to promote low carbon energy options and thus reduce emissions in the energy sectors. However, laws that subsidise and encourage the use of fossil fuel could serve as barriers to departing from these traditional fuel sources, thus extending the oil and gas era in Sub Saharan Africa. The clean energy transition in line with the objectives of the Paris Agreement may be delayed as a result of the global community's recent interest in oil and gas from the region following the Russian-Ukrainian conflict. Further, the absence of laws and regulations to promote the use of renewable, low-carbon energy sources is an added barrier to moving towards net zero. Additionally, poor citizen enlightenment and stakeholder engagement are challenges to achieving its target of keeping the planet's temperature below 2°C compared with pre-industrial levels. So far, achieving this goal has been an ordeal for the global community, as countries have different national, social, economic and political priorities and circumstances. Thus, the disposition per country to address climate change is hinged on national sovereignty, national circumstances and capabilities.

A review of the portfolio of major emitting countries shows that they are also the most industrialised countries compared with countries at the bottom of the ladder regarding their carbon emissions. The energy sector is the source of about three-quarters of global emissions; thus, addressing carbon emissions in the energy sector is essential for GHG mitigation. Significantly, major emitting countries also have carbon-intensive energy industries powered by coal, gas and oil. Countries in Africa, particularly SSA, generate the lowest carbon emissions, have the lowest capacity for energy generation, and have the lowest development indices. These national circumstances, common amongst SSA countries, raise the question of the implication of a 'race to net zero' for countries whose emissions are already close to zero or around 1%. It also raises the question; should countries with low carbon emissions, low energy access rates, and high poverty indices also be racing to net zero? What impacts could laws and policies pushing for climate change mitigation and net zero have on the societies in these countries? These questions are further relevant in light of the guiding principles of the UNFCCC, i.e. the principles of common but differentiated responsibilities and respective capabilities (CBDR-RC), prevention, and the polluter pays.

Sub-Saharan Africa is a region with varied socio-political systems, legal frameworks, and socio-economic conditions. Examining selected countries within the sub-region allows for a deeper understanding of the unique challenges and opportunities faced in implementing climate change policies, ultimately contributing to effective climate action and sustainable development.

Table 1. Development Indices for the DRC, Kenya, and Nigeria

Country	Population (mill) 2021	GDP (current US$ billion) 2021	Access to electricity (% of population) 2020	CO2 emissions (metric tons per capita) 2020	Population growth (annual %) 2021	GDP growth (annual %) 2021
DRC	92.377986	53.96	19.1	0.03	3.10	5.71
Kenya	54.985702	110.35	71.438	0.42	2.23	7.52
Nigeria	211.400704	440.78	55.4	0.57	2.52	3.65

Compiled from: the World Bank, 2022

Table 1 highlights social, economic and environmental factors in the DRC, Kenya, and Nigeria. The above socio-economic characteristics indicate that the SSA region struggles with challenges synonymous with most developing countries across other continents recovering from the post-COVID-19 recession. (World Bank, 2022). The climate across the region has not fared better than the economy; GHG emissions are still minimal but witnessed a slight increase after the Covid-19 pandemic, although a handful of SSA countries have managed to slow down their emission rates (Botswana, Namibia, South Sudan, The Gambia and Equatorial Guinea (World Bank, 2022).

For countries in SSA struggling to attain industrialisation, climate justice, equity, and the overriding need for development cannot be ignored. With an estimated population of 1.7 billion (World Bank, 2022), 35% of whom are without electricity and have a life expectancy below 65 years, the need for development is paramount. (Corfee-Morlot et al., 2022).

Countries in SSA still grapple with providing their people with social amenities and critical infrastructure. They also face burning national issues, including peacekeeping, pandemics, and global inflation. Added to this is the challenge of climate change and its impacts. A major negative impact of climate change on people's livelihoods in SSA is its impact on crop production (Jafino et al., 2020, USAID, 2021), as climate change directly translates to shrinking water bodies and reduced access to water resources due to droughts. These negative impacts added pressure on already strained inter-human, inter-tribe relations, leading to increased conflict in conflict-prone areas (IPCC, 2021).

Recent studies have emphasised the benefits and co-benefits of climate actions in assisting SSA countries in achieving sustainable development. (Afaha & Ifarajimi, 2021, Akinbusoye, 2022). For example, deploying low-carbon and clean energy sources for fuel could create employment opportunities, support a cleaner environment, promote economic diversification, reduce maternal mortality, and promote energy access rates. On the other hand, scholars have identified that initiatives in line with climate action could result in human rights violations, the displacement of vulnerable people and poverty exacerbation. (Olawuyi, 2016, Adeola & Viljoen, 2022). Policies and laws to address climate change could also lead to the sacking of indigenous populations who rely on the forests in the Sub-Saharan African region (Koné, 2022). This study considered the socio-legal impacts of climate change laws and policies on the people and societies in the DRC, Kenya, and Nigeria.

The study examined socio-legal opportunities and barriers to achieving the objectives of the Paris Agreement in three countries within SSA which are at varying stages of industrialisation. Socio-legal studies examine the interactions of law and society to understand the law's impacts on society (Bradshaw, 1997). Hence, the study examined the relationship between climate change laws and related social and legal circumstances in the DRC, Kenya and Nigeria. It highlights socio-legal challenges faced in adapting the main objectives of the Paris Agreement to societal circumstances. These countries represent the South, West and East areas in SSA and are a heterogeneous mix of SSA countries with varying socio-economic characteristics. The countries also serve as exemplary case studies for an analysis of the interface between existing climate policies and social and legal circumstances (Marigi, 2017). Additionally, these countries frequently deal with risks resulting from climate change, such as droughts, and floods. The effects are exacerbated by deep rural poverty, limited government capacity, and exposure to additional distress (Guillaumont & Simonet, 2011)

The study highlights successful initiatives and best practices that have emerged from the region, showcasing the resilience and determination of Sub-Saharan African countries in addressing climate change. Also, it critically analyses the gaps and areas requiring urgent attention to accelerate progress in the global fight against climate change. It examines how climate-related laws seeking to promote climate change mitigation action play out in social and economic contexts in the DRC, Kenya and Nigeria. It does this by examining the nationally determined contributions of countries and climate change-related laws and policies in the light of their overall national circumstances. The chapter expresses the view that countries in SSA should avoid the pressure from the global North to join the 'race to net zero' but rather ensure a gradual, practical energy transition in line with national circumstances and peculiarities. There is a need for countries in SSA to explore nationally appropriate opportunities and put forward indigenous solutions in line with their national circumstances as their contribution to the climate agenda. The chapter recommends that developing countries pay more attention to the solutions and roles they can proffer in line with the global climate agenda according to the CBDR-RC principle. Such solutions should be hinged on an approach that puts citizens at the centre of climate action.

BACKGROUND

Paris Agreement: Core Objectives in the Sub-Saharan African Context

The Paris Agreement, adopted in 2015, marked a monumental global effort to combat climate change and mitigate its devastating impacts. It set ambitious goals to limit global warming to well below 2 degrees Celsius above pre-industrial levels and pursue efforts to limit the temperature increase to 1.5 degrees Celsius. While the agreement serves as a unifying force, its implementation remains a complex challenge, particularly in Sub-Saharan Africa.

The global community adopted the Paris Agreement to build on the progress made under the erstwhile Kyoto Protocol of the UNFCCC. The modus operandi of the 2005 Kyoto Protocol was heavily hinged on the CBD-RC principle, as the developed countries that were signatories carried the burden of mandatory climate action. The kick against this approach led to negotiations for a new climate agreement. The Paris Agreement aims to achieve more than the Kyoto Protocol by inviting all country parties to actively respond effectively and progressively to the common concern of climate change through their nationally determined contributions (NDCs). There is hope that through the Paris Agreement, a more collective bottom-up approach to climate action will help address global warming; however, with seven years of the Paris Agreement left, more progress has yet to be achieved (Maizland, L. 2021).

The 2015 Paris Agreement represents a global consensus to curb climate change and allow nature to begin its long but steady journey to rejuvenation. It presents two levels to addressing climate change: the NDC country-level of governance and the supervisory level spearheaded by the United Nations Framework Convention on Climate Change (UNFCCC). The UNFCCC monitors benchmarks fixed by countries in line with their plans, policies, and contributions to global decarbonisation and means of temperature reduction (AfDB, 2016). The essence of countries designing their NDCs is to incorporate their peculiarities and circumstances to increase the feasibility of meeting set targets and deliverables. The Paris Agreement has been signed by 48 SSA countries, 47 of which have ratified the agreement and submitted their NDCs, except for Libya. (UNECA, 2022) Laudable as the commitments in the respective NDCs are, fulfilling their promises has been challenging due to societal and political circumstances (Paterson et al., 2019).

Given their peculiar circumstances, SSA countries must re-evaluate their roles in the global fight against climate change. Additionally, climate change-related projects require societal acceptance and understanding of the issues in order to harness indigenous, homegrown solutions as well as national and regional buy-in and funding. Benin, Kenya, Nigeria, Uganda, and Mauritius have enacted climate change laws within the SSA countries. These laws are structured as framework laws to guide future legal and political efforts in mainstreaming climate action across national sectors. However, the laws are embryonic and partially implemented.

In line with the provisions of the UNFCCC, the Paris Agreement is premised upon the principles of equity and justice, the CBDR-RC principle, the precautionary principle, and the polluter pays principle. Thus, developed country parties should take the lead in climate action. Also, developing countries' specific needs and circumstances, particularly those most vulnerable to climate change impacts, must be considered in determining and pursuing their NDCs.

The CBDR-RC principle of the UNFCCC concedes countries' different capacities and responsibilities in their response to climate change. The principle appreciates the common problem of climate change, its historical antecedence, and countries' national and regional development differences.

Thus, governments must act in line with the principles of equity and follow their common but differentiated responsibilities and respective capabilities. Articles 3 and 4 of the UNFCCC recognise this principle and state the various roles developed and developing country parties should play to balance the climate. The CBDR-RC principle is retained and reflected throughout the Paris Agreement (Adejonwo, 2018).

Thus, regarding the roles to be played in the global climate agenda, developed country parties must take the lead in combating climate change. Article 3 (1) UNFCCC. This leadership role involves financial and technical assistance to developing country parties. In line with this, Article 9 of the Paris Agreement mandates developed countries to provide financial resources to assist developing countries in climate change mitigation and adaptation in line with the polluter pays principle in environmental law. On the other hand, all countries are expected to implement measures to mitigate climate change by identifying and managing emissions from their sources and removing emissions through carbon sinks Article 4, (1b) UNFCCC. Under the Paris Agreement, these measures are to be determined by countries in their nationally determined contributions. It is worth noting that Article 4 (3 & 4) of the Paris Agreement emphasises that country NDCs should reflect their national circumstances.

While the Paris Agreement recognises the contributions of developed countries to the current climate crisis and thus the common but differentiated responsibility principle, the net-zero agenda is gaining popularity.

Countries in SSA are aligned with the Paris Agreement, as evidenced in their NDCs. While Africa's climate implementation record has been criticised (Munang & Mgendi, 2016), the continent is determined to follow the Paris Agreement. However, challenges with the implementation of laws and regulations, human rights violations, limited financial capacity, and other national circumstances constrain the ability of countries in SSA to achieve global climate goals. Furthermore, these countries have clarified in their NDCs that attaining their goals and targets toward net zero is contingent upon financial support from developed countries.

The core objectives of the Paris Agreement are highly relevant to the context of Sub-Saharan Africa. The region is particularly vulnerable to the impacts of climate change, with its countries facing challenges such as food insecurity, water scarcity, extreme weather events, and ecosystem degradation (Adesete A. A. et al., 2022). These challenges, coupled with limited adaptive capacity and socio-economic disparities, make it crucial for Sub-Saharan African countries to prioritise climate action and align their efforts with the objectives of the Paris Agreement. The core objectives of the Paris Agreement in the Sub-Saharan African context include mitigation, adaptation, finance and technology transfer.

The 27th Conference of the Parties (COP27) to the UNFCCC provided an opportunity for global leaders, policymakers, and stakeholders to review progress in implementing the Paris Agreement and strengthen international cooperation on climate action. The 2022 COP 27 witnessed the establishment of the Loss and Damage fund which is intended to address damage in countries that have suffered the most loss and damage from events induced by climate change. The fund presents an opportunity for African countries to be compensated by developed countries for their role in the current climate situation. However, experiences from the failed 100-billion-dollar annual donation promised since 2009 by developed countries at COP 15 in Copenhagen for climate finance casts a shadow on the genuineness of financial commitments made by developed countries.

COUNTRY CASE STUDIES

The Democratic Republic of Congo (DRC) (National Circumstances)

The DRC is a signatory to the UNFCCC, the erstwhile Kyoto Protocol and the Paris Agreement. The country is committed to climate change mitigation and adaptation per Article 4(1) of the Paris Agreement.

GHG emissions from the DRC are majorly from forest and land use, waste management, agriculture and the energy sectors. Due to the country's low level of development, its total contribution to global emissions is relatively low, at about 0.47%. (Figure 1 shows the GHG emission profile for the DRC). Despite its low GHG emission status, carbon emissions from the Forestry and Other Land Use and energy sectors are of concern. Though these emissions are quite minimal, these two sectors are thus amongst the major sectors targeted for mitigation.

The DRC is yet to pass a climate change law; however, its 2021 NDC indicates that a framework climate change law may be on the horizon. The law is expected to establish an institutional framework for addressing climate change and promote intersectoral coordination in addressing climate change concerns. The DRC faces extreme vulnerability to the effects of climate change as extreme weather events such as violent rains, floods, erosion, lengthened dry seasons, and heat waves threaten the primarily agrarian means of livelihood and homes. Hence adaptation to the effects of climate change is an issue as the DRC must strengthen its resilience to climate change as contained in the country's National Action Program for Adaptation to Climate Change (PANA, 2006).

The DRC continues to endure political instability, and decades of civil war have plunged the country into dire poverty and underdevelopment. Its economic situation has further suffered contraction due to the Covid -19 pandemic, which restricted commercial activities. Thus, poverty is a major challenge for the people of the DRC, many of whom rely on forests for their livelihood and energy needs.

The DRC has committed to a 19% national GHG reduction target with conditions and an unconditional target of 2% (NDC). A sum of USD 25.60 Billion is needed for its mitigation initiatives. The primary mitigation measures to be implemented are targeted at reducing emissions from deforestation, forest degradation and other forest conservation efforts (REDD+), as illegally exploiting the forests for fuelwood encourages deforestation, forest degradation and environmental pollution in the DRC.

While it is acknowledged that there is a need to preserve forests and prevent deforestation, there is also a need to recognise that the people in the DRC are deeply connected to and rely heavily on forests and farmlands for their livelihood and socio-cultural practices. (Brown, 2011). Thus, REDD+ efforts must seek to balance the efforts towards climate change mitigation against the socio-cultural rights of indigenous people to property, development, and their cultural and religious practices as decided by the African Commission on Human and People's Rights in the case of the Centre for Minority Rights Development (Kenya) and Minority Rights Group International on behalf of Endorois Welfare Council v. Kenya.

Furthermore, due to competing economic interests and limited financial resources, international donors and development partners are needed to finance the implementation of the NDC as the country is mainly concerned with its national development priorities, which cover health, social, economic, infrastructural issues and poverty alleviation. Financing its NDC has, therefore, not been a priority for the DRC.

Figure 1. GHG emission profile for DRC
Source: DRC's NDC, 2021

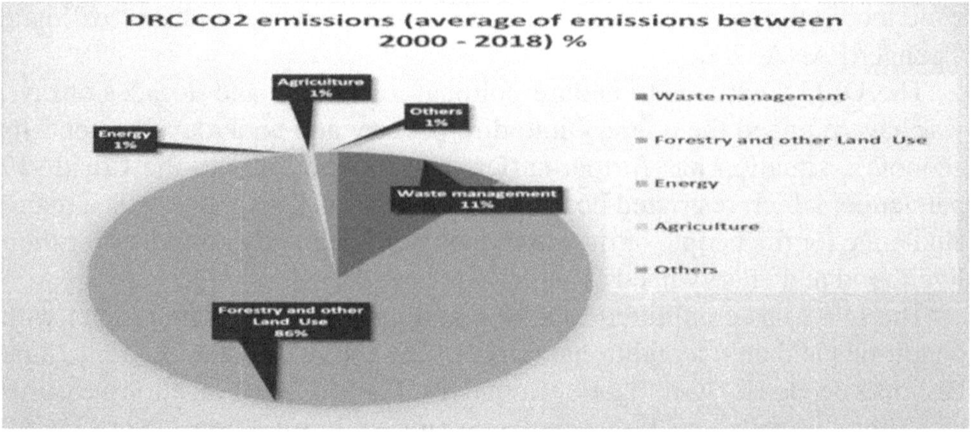

Socio-Legal Opportunities and Barriers to Climate Change Mitigation in the DRC

The vast forest reserves in the Congo Basin represent ~10% of the global tropical forest. As a tropical forest country, the forests in the DRC play a significant role as a carbon sink. The rich forest areas in the DRC play a massive role in eliminating carbon pumped into the air by major emitting countries. Thus, the forests provide a huge opportunity for the DRC to play a vibrant role and trade in the global carbon market. Under the implementation of Article 6 of the Paris Agreement, the country can sell carbon credits on the international carbon market in exchange for renewable energy technology or mitigation or adaptation projects.

A significant challenge for the government of the DRC is balancing the DRC's climate change policies which promote REDD+ mitigation efforts, and the rights of indigenous men and women who rely on the forests for their energy, health, cultural and religious practices, and as a source of livelihood. Koné notes that the REDD + strategy deployed in the Mai Ndombe province was implemented without the involvement and consent of the local community, resulting in destabilisation and exacerbated conflicts in the area (Koné, 2022) . Indigenous people play a central role in forest preservation and sustainable management of natural resources. However, their skills and knowledge are unrecognised (Brown, 2011). Indigenous people must be better represented in decision-making processes relating to the country's development and national and international issues such as climate change and poverty eradication.

Indigenous people, particularly women, can play a major role in providing nature based solutions in the context of climate action, while also helping to promote food security. In order to achieve this, the government of DRC must collaborate with the indigenous people and leverage their indigenous knowledge in forest management in line with its REDD+ ambitions.

Additionally, the DRC's NDC seeks to improve the population's access rate to electricity, which is currently very low. The 2005 Constitution of the DRC guarantees the right of citizens to access electricity, but this promise is yet to be fulfilled. A significant barrier to achieving the vision is the country's need for more financial resources and investors in the energy sector.

Investors must invest more in clean energy solutions to address local energy needs and provide available, accessible, affordable alternative energy. The current low electrification rate presents an opportunity to deploy renewable energy solutions, develop technical capacity, and train youth to design and create clean energy technologies, further providing job opportunities.

Law No. 14/011 (Electricity Sector) regulating the energy sector was passed in 2014 to liberalise the energy sector, promote reliable access to energy to unreached areas and guarantee an improvement in competition for energy operators. The law covers the transmission and distribution sectors. However, to date, there are fundamental gaps in power generation, transmission and distribution due to the poor management of the sector. Overall, the government needs to promote all sources of renewable energy.

The DRC's electrification-related mitigation activities are estimated at USD 2.23 Billion. The target is to achieve an increased electrification rate in rural, peri-urban, and urban areas through renewable energy and improve the share of renewable energy in the country's energy mix by 2030. The DRC's electricity sector challenges include limited institutional capacity, poor equipment management due to low technical skills and poor maintenance culture. Existing hydropower plants supplying electricity to major cities need to perform better and major maintenance. Limited financial capacity underscores the challenges limiting the development of big hydropower stations and the building of new power lines.

The DRC's inga dams provide great potential for hydroelectricity and align perfectly with the country's energy targets under its NDC. The dams provide an opportunity to provide up to 40 GW of low carbon, clean, hydro-powered energy to the people of the DRC and for export to neighbouring countries. The project has the potential to double Africa's energy capacity and increase the standards of living for the people in the DRC and throughout Africa. Plans are underway for a combined great inga dam project. However, these plans have stalled for decades due to financial constraints, human rights concerns and fears of corruption by the country's leaders. Congolese activists have argued that available financial resources should be channelled to restore the existing Inga Dams 1 and 2, which operate far below capacity and focus on mini-grids that will serve the people's needs. The project is also said to be unjustified by the DRC's current rate of energy demand. These concerns amplify the voices of indigenous men and women displaced during the building of Inga 1 and 2 (in 1972 and 1982, respectively), who were neither compensated nor resettled and had no access to the electricity generated from the dams. Indigenous people in the DRC also claim that water, energy, and other natural resources are exploited by the government and private mining companies primarily for the benefit of private interests.

The International Climate regime can be a vehicle to promote the use of renewable energy to support rural and urban electrification, address the energy poverty situation, and promote low-carbon development in the DRC. However, in its efforts to adopt a low-carbon development model, the DRC balances competing legal and economic interests. The government must consider the rights and needs of indigenous people, who need to be recognised as stakeholders and knowledge holders, not victims (AIPP, 2021).

Kenya (National Circumstances)

In East Africa, Kenya is home to the second largest population, 47.6 million by 2019 and projected to reach 60.4 million by 2030 (the Republic of Kenya, updated NDC, 2020). Kenya boasts an economy which is large and diversified. The arid nature of most Kenyan land predisposes the country to major environmental risks such as floods and drought, which are significant climate hazards affecting lives and livelihoods negatively. The Kenyan economy mainly depends on agriculture, water, energy, tourism and wildlife, and climate change highly impacts these sectors. Kenya is also a signatory to the UNFCCC, the erstwhile Kyoto Protocol and the Paris Agreement.

The country has a mix of renewable energy sources such as solar, wind, geothermal and hydro for electricity generation. However, despite these resources, over half of the country's population uses firewood for cooking. Electricity is generated mainly through hydropower, geothermal energy, diesel, and some nominal wind generation. Even though the country has coal reserves that can be exploited for electricity generation, using coal for electricity is not widespread because of its significant environmental impacts (Kenya NDC, 2020). Kenya can either use her coal reserves and impact the environment negatively or forgo this resource entirely. Forgoing this resource implies a heavy dependence on international support to develop cleaner energy sources.

As of 2015, Kenya's emissions were at 93.7MtCO2e and are projected to increase to 143MtCO2e by 2030. Figure 2 shows Kenya's GHG emission profile. Agriculture is the major contributor to Kenya's emissions, contributing ~40% due to general fermentation processes. Land Use Land Change and Forestry contributed 38%, while Industrial Processes & Product Use and Waste Management contributed 3% and 1%, respectively. By 2030, energy could be the most significant contributor due to increased fossil fuel consumption from electricity generation, heating for the industrial, commercial and domestic sectors, and transport.

Figure 2. GHG emission profile for Kenya
Source: KENYA NDC 2020

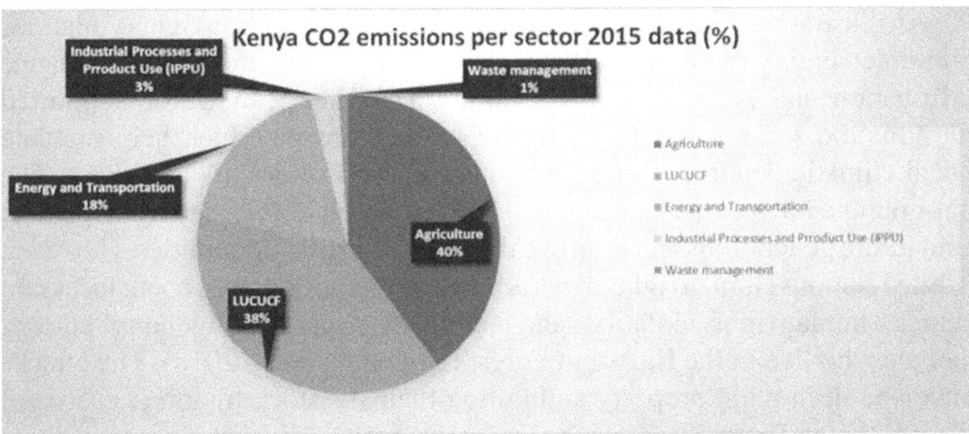

Climate change has posed an obstacle to development efforts in Kenya, with the country losing up to 3-5% of its Gross Domestic Product (GDP) annually due to the impacts of climate change on the economy.

Kenya plays a leadership role in climate change mitigation and adaptation in SSA. The government has implemented laws and policies to adapt to and mitigate climate change. Kenya was the first country to pass a climate change law in Africa (Climate Change Act, 2016). Additionally, its Climate Change Policy, Climate Change Response Strategy, National Action Plan, and National Adaptation Plan provide a framework for climate action. The overall objective is to develop Kenya's low-carbon emission development path.

Section 42 of The Constitution of Kenya (2010) mandates the government of Kenya to ensure every Kenyan has a clean and healthy environment formulating climate change legislation and institutions that seek to restrict access to forests and steer the country towards sustainable development, human well-being and green growth. However, this right must be balanced with the rights of Kenyans to their ancestral lands and cultural practices recognised in Section 56 of the Constitution.

Socio-Legal Opportunities and Barriers to Climate Change Mitigation in Kenya

Many of the climate mitigation projects ongoing in Kenya are ecosystem-based and situated in the forests and open lands. There is a risk of a contradiction between climate change policies seeking to mitigate carbon emissions and the realisation of the rights of indigenous people who depend on the forests for their food, water and livelihood.

An illustration of such a violation is the challenge of human rights violations in implementing the Kenyan Water Towers Protection and Climate Change Mitigation and Adaptation Programme. The European Union-supported Programme seeks to conserve Kenya's water towers which help regulate local climatic conditions and function as carbon reservoirs and sinks. The programme is expected to preserve the environment for the benefit of present and future generations and improve the livelihood of communities. However, United Nations Human rights experts have expressed concerns about incessant attacks, human rights violations and forceful evictions of indigenous Sengwer people who live in the Embobut Forest (United Nations, 2018). The attacks involved destroying property and killing their livestock by forest guards of Kenya Water Towers.

Indigenous people, particularly women, are often marginalised in decision-making processes in the DRC. Action and decisions to mitigate climate change per the Paris Agreement must be taken inclusively. Indigenous women are invaluable in the fight against climate change. Their traditional knowledge of ancient forest practices could be helpful in forest and biodiversity preservation and promote sustainable use of natural resources.

Thus, the UNDP recognises that "National climate policies must recognise Indigenous Peoples as rights-holders, knowledge-holders, and equal partners in efforts to combat climate change" (UNDP, 2022). Kenyan climate change laws and policies must be implemented to reflect an appreciation of indigenous people's rights to life, property and land security.

Nigeria (National Circumstances)

Home to over 200 million people, Nigeria is Africa's largest nation with the largest economy. Reliance on oil and gas export revenues renders the economy volatile due to constant oil price fluctuations and global economic shocks. Nigeria's land and water resources are at severe risk due to the impacts of climate change. Droughts have displaced pastoral farmers in Northern Nigeria. Flooding and erosion destroy the assets and property of farmers in the rainforests and coastal zones, hence the need for more attention on adaptation to climate change. While Nigeria has committed to the goals of the international climate change framework to stabilise GHGs, with a rapidly growing population, the government identifies its top priorities to include accelerated development and achieving food security and poverty eradication.

The Nigerian government faces a major challenge of electrification and energy poverty. (Nigeria's NDC, 2021). According to the World Bank (2021), 85 million Nigerians lack access to electricity from the national grid. Nigeria ranks 131 out of 190 countries in access to electricity. This constraint significantly hinders growth in the private sector (World Bank Doing Business Report, 2020). This challenge poses a barrier to fulfilling the mandates of the international climate change regime in some ways, as it presents the issue of conflicting legal and social priorities. For example, the country can continue exploring available oil and gas reserves for income and energy or channel funds to develop renewable energy technology and invest more in renewable energy sources. Though an attractive option, the latter is more advanced and complex as the country is renowned in the global oil and gas market. At present, the government is pursuing both options. However, the oil and gas industry is dominant.

Nigeria has been active in the UNFCCC since its inception by ratifying the Kyoto Protocol in 2004 and the Paris Agreement in 2017.

Currently, GHG emissions from Nigeria are estimated at 1%, with the majority coming from its energy sector (up to 60%) (Nigeria NDC, 2021). Fugitive emissions from oil and gas make up 36% of emissions from the energy sector. Figure 3 shows Nigeria's GHG emission profile. Nigeria's mitigation efforts, therefore, lean heavily on reducing carbon emissions from its energy sector. Strategies and plans in line with this include promoting energy efficiency and access to clean energy, reducing emissions from gas flaring, using flared gas, and using liquified petroleum gas (LPG) and other clean cooking energy options. These strategies are supported in official policy documents such as the Rural Electrification Policy (REP, 2005), Renewable Energy Master Plan (REMP, 2012), National Energy Policy (NEP, 2015), and the Rural Electrification Strategy and Implementation Plan (RESIP, 2016), amongst others.

While the country has immense potential and opportunities to develop its renewable energy sources, inconsistencies between policies and laws in the energy and power sector hinder rapid progress in this area, thus hindering ambitious efforts. For example, the government's target to achieve 75% of rural electrification by 2020 still needs to be reached.

With a major energy crisis on hand, a significant challenge for the country is juxtaposing its self-imposed mitigation commitments under its NDC with the pressing need to expand its energy resource base to produce enough energy to meet the growing demand. A similar challenge lies in defining and pursuing its energy transition path to ensure maximum benefits from all available resources, including forests and oil and gas.

Nigeria's Climate Change Act, (CCA) 2021 sets the target time frame for the country to achieve net-zero carbon emission as 2050-2070. The Act's provisions apply to private and public sector players and covers all sectors of the economy. The law was passed to demonstrate commitment to the UNFCCC and the Paris Agreement. The Act provides a framework for mainstreaming climate change actions, provides a system for carbon budgeting, and establishes the National Council on Climate Change (NCCC). Section 3 of the CCA 2021, established the NCCC with the powers to coordinate the implementation of sectoral targets and produce guidelines for the regulation of GHG emissions. The council is also responsible for overseeing the implementation of the National Climate Action Plan and administering the Climate Change Fund established under the Act. It is hoped that as the Act moves on to implementation, the NCCC will ensure the mainstreaming of climate change into the national development plans and programmes; formulate policies and programmes relating to climate change and involve the citizenry in climate change planning, projects policy making.

The Act, in section 15 (2) (e, f, i, and j) highlights that funds from the established Climate Change Fund shall be applied towards climate change advocacy and information dissemination. This is crucial and must be implemented to promote effective and just climate governance. While the fund is to be applied towards funding innovative climate change mitigation and adaptation projects, the fund is also to be used in conducting assessments of climate change impact on vulnerable communities and population. Additionally, the Act states that the funds will be used in incentivizing private and public entities for their efforts towards transiting to clean energy and sustaining a reduction in GHG emissions. It does not however state whether such incentives and benefits will apply to communities involved in GHG emission reduction activities. There should be additional considerations for incentivising community and individual efforts in their efforts towards GHG reduction as indigenous people who have been involved in forest conservation activities since time immemorial may be left out. The Act also does not clearly address the possibility of mitigation or adaptation projects negatively impacting vulnerable groups or communities. It is hoped that as the NCCC commences its duties, these issues will be addressed to ensure that there is adequate protection and compensation for vulnerable groups.

Figure 3. GHG emission profile for Nigeria
Source: NIGERIA NDC 2021

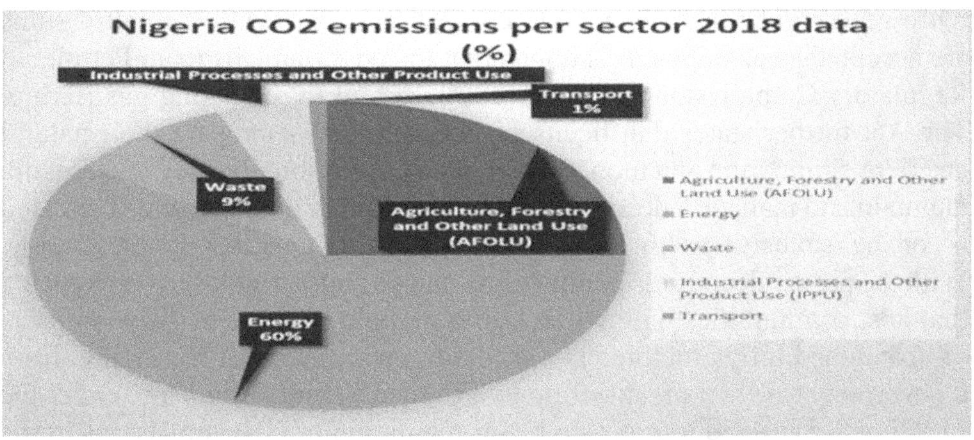

Socio-Legal Opportunities and Barriers to Climate Change Mitigation in Nigeria

The federal government's 2021 energy transition plan involves increasing access of Nigerians to modern energy services to address poverty and climate change and drive economic growth. The plan emphasises the need for a just transition that preserves the jobs and livelihoods of people currently dependent on the fossil fuel industry. Several initiatives in Nigeria's energy sector contribute to climate change mitigation.

A major strategy in line with the Paris Agreement in the oil and gas sector is the plan to promote gas as a cleaner alternative to oil (Nigeria NDC, 2021). Natural gas has thus been identified as the nation's transition fuel, and the period between 2020 and 2030 has been tagged the decade of gas. Therefore, the newly introduced Petroleum Industry Act (PIA) 2021, aims to encourage the development of natural gas infrastructure to create a vibrant domestic market for Nigeria's abundant gas resources. Though laudable as gas is a cleaner fuel than oil, socio-economic and infrastructural limitations such as Nigeria's limited gas pipeline distribution network significantly hinder the government's aspiration. The menace of gas pipeline vandalism also constitutes a challenge to the gas agenda (Oke, 2021).

In line with the country's net zero emission target, Section 104 of the PIA 2021 highlights that penalties shall be paid by licensees for flaring natural gas. The Act further prohibits the flaring or venting of natural gas in Section 105 (1), and elucidates that licensees or lessees shall pay penalties subject to the provisions of the Flare gas (Prevention of Waste and Pollution) Regulations. Penalties have always existed for gas flaring, but the challenge has been with weak regulatory powers, and corruption in the sector as paltry sums are accepted as penalties. It is hoped that the Nigerian Upstream Petroleum Regulatory Commission will be more successful in regulating gas flaring. The Act further states that licensees or lessees must prepare their natural gas flare elimination and monetization plans. Additionally, they must set up, maintain and manage a decommissioning and abandonment fund. This fund is for the exclusive payment of decommissioning and abandonment costs. With the proper decommissioning of oil and gas infrastructure, it is expected that host communities will not be left to grapple with the negative impacts of abandoned infrastructure. The recent transformation of the NNPC from a government-owned organization to a limited liability company, under the PIA is also expected to promote better regulation of GHG emissions in the Nigerian energy sector as the government should no longer be conflicted in playing the dual role of regulator and industry player. These initiatives align with the international climate agenda as contained in the Paris Agreement and they aim to reduce the carbon footprint in the Nigerian energy sector, in line with the decarbonization strategy enshrined in the Nigerian Energy Transition Plan (NETP) for the Nigerian oil and gas industry.

Another major step in line with the low carbon development objective of the Paris Agreement and Nigeria's climate change mitigation efforts is the removal of the subsidy for petroleum products. Deregulation of petroleum product pricing is one of the ways to ensure that the actual cost is reflected. A reflection of actual costs, including environmental costs, is expected to place petroleum products on a level playing field with renewable energy sources, thus allowing consumers to make more informed choices. Additionally, actual costs ensure that polluters pay for environmental damage, if any, to the environment. The federal government is therefore poised to follow through with the removal of subsidies on petroleum products. While these efforts are laudable, there are kickbacks from stakeholders who argue that subsidy removal on petroleum products is negatively impacting over 94 million Nigerian citizens living below the poverty line.

NIgerias NDC sheds light on the government's plan to lift 25 million households from using carbon-intensive charcoal and fuel wood to LPG, a cleaner cooking fuel. The strategy is also aimed at simultaneously addressing deforestation and forest preservation. However, recent increases in the price of LPG due to deregulation have resulted in women returning to firewood and charcoal for their cooking and heating needs. Additionally, studies have shown that some prefer cultural and traditional fuelwood for cooking due to habit and taste preferences, underscoring the need for increased awareness and public education about charcoal cooking and deforestation's environmental and health impacts. The government's efforts to provide improved cookstoves to indigenous women and reduce their reliance on firewood and charcoal may be futile if traditional women cannot access reliable, affordable and suitable alternatives.

The government's ongoing Solar Power Naija programme which the Rural Electrification Agency is anchoring aims to expand energy access to 25 million individuals by providing solar home systems or mini-grid connections. There are immense benefits in powering rural off-grid areas, schools and health centres with solar power; however, there are financial constraints in promoting the reach of solar power due to the high costs of solar-powered mini-grids. Additionally, the absence of a law to create a stable framework for renewable energy deployment in Nigeria implies that government policies may change with a change in government administration. Scholars have highlighted the discrepancies in Nigeria's energy policy objectives and the legal realities in Nigeria. Oke argues that the absence of renewable energy law is a significant deterrent factor in propelling the growth of cleaner, low-carbon sources in Nigeria (Oke, 2021). Further, the existing Electric Power Sector Reform Act 2005, the energy sector's guiding law, needs to provide for renewable energy sufficiently.

It is expedient that the government manages Nigeria's energy transition creatively to ensure the protection of the majority of the Nigerian populace who live below the poverty line from falling deeper into abject poverty. The recently passed Climate Change Act will enhance the mainstreaming of climate action in Nigeria across relevant sectors, including the energy and power sectors.

SOLUTIONS AND RECOMMENDATIONS

The Paris Agreement has heralded an era of increased country participation. SSA countries have much to contribute to the objectives of the Paris Agreement, particularly in line with climate change mitigation. Their forests serve as lungs to absorb carbon emitted by the world's developed countries. The region's rich abundance of natural energy resources also positions SSA countries at a vantage point in addressing climate change and all forms of poverty. Thus the teeming youth, indigenous populations and the rich forests in SSA should play more central roles in the solution to climate change as these are assets that can be developed and leveraged in international climate change negotiations. Countries in SSA and Africa must prioritise their national development goals in line with sustainable development models. They must resist the pressure to do away with their fossil resources, as only a balanced approach to energy can solve the region's challenge of energy poverty.

In pursuing their NDCs, developing countries in SSA must take deeper consideration of the needs of their citizens and their national and social circumstances. They must take a firm stand on the need to protect indigenous people's rights and lands and ensure that climate change mitigation projects are conducted inclusively, considering the socio-economic circumstances of their people. The pressure from international funders and the international community at large to phase out fossil fuel may not be fully suitable for SSA countries that are yet to reap industrialisation benefits from developing their natural resources.

FUTURE RESEARCH DIRECTIONS

This chapter examined the implications of the GHG mitigation objective of the Paris Agreement and related laws, implemented projects and policies in the DRC, Kenya and Nigeria. There is a need to investigate the impacts of climate change mitigation laws and policies on other countries in SSA. Although the global climate agenda focuses more on mitigation than adaptation, another researchable area is the investigation into the Paris Agreement's adaptation objective and its effectiveness in achieving its aim to support climate change adaptation in developing countries.

CONCLUSION

The reality of Climate change is no longer foreign to countries in Africa, and SSA countries have thus improved their activity levels in the fight against climate change. Consequently, developing countries are no longer asking 'what should be done? However, 'What can we do about climate change? Despite the current disposition of countries in SSA to be part of the global fight against climate change, the historical antecedent of climate change must be remembered. African countries must continue to emphasise their current circumstances and the need to solve their peculiar problems by using their natural resources and managing them appropriately.

This chapter finds that due to the socio-economic circumstances in the DRC, Kenya and Nigeria, there are challenges and opportunities in implementing the GHG mitigation objectives of the Paris Agreement. The differing national circumstances of each country show that there is no one size fits all solution to addressing climate change. Developing countries in SSA must therefore own their development process and carefully but responsibly plan their energy transition trajectories to benefit their citizens while seeking to comply with international climate change laws.

ACKNOWLEDGMENT

This research received no specific grant from any funding agency in the public, commercial, or not-for-profit sectors.

REFERENCES

Adejonwo, O. (2018). Nigeria's commitments under the climate change Paris Agreement: legislative and regulatory imperatives towards ensuring sustainable development. *4th Scientific Conference of the Association of Environmental Law Lecturers from African Universities in cooperation with the Climate Policy and Energy Security Programme for Sub-Saharan Africa of the Konrad- Adenauer-Stiftung and UN Environment*. UN Environment.

Adesete, A. A., Olanubi, O. E., & Dauda, R. O. (2022). Climate change and food security in selected Sub-Saharan African Countries. *Environment, Development and Sustainability*. doi:10.100710668-022-02681-0 PMID:36186913

Afaha, S. J., & Ifarajimi, G. D. (2021). Energy poverty, climate change and economic growth. *African Journal of Economic and Sustainable Development*, *4*, 98–115. doi:10.52589/AJESD-U3LCOY0P

Awaworyi, C. S., Smyth, R., & Trong-Anh, T. (2022). *Energy poverty, temperature and climate change*. Research Gate. https://www.researchgate.net/publication/360001563_Energy_poverty_temperature_and_climate_change

Ayoade, M. (2018). *Bridging the gap between climate change and energy policy options: What next for Nigeria? 4th Scientific Conference of the Association of Environmental Law Lecturers from African Universities in cooperation with the Climate Policy and Energy Security Programme for Sub-Saharan Africa of the Konrad- Adenauer-Stiftung and UN Environment*. UN Environment.

Bijoy, C. R., Chakma, A., Guillao, J. A., Hien, B., Indrarto, G. B., Lim, T., Min, N. E., Rai, T. B., Smith, O. A., Rattanakrajangsri, K., Thuy, H. X., & Zunga, U. (2022). *Nationally Determined Contributions in Asai: Do governments recognise the rights, roles and contributions of Indigenous Peoples?* Asia Indigenous People Pact Report. https://aippnet.org/nationally-determined-contribution-asia-governments-recognizing-rights-roles-contributions-indigenous-peoples/

Bradshaw, A. (1997). Sense and sensibility: debates and developments in socio-legal research methods. In P. A. Thomas (Ed.), *Socio-legal studies* (pp. 99–122). Dartmouth Publishing Company. https://link.springer.com/article/10.1007/s10991-010-9069-6

Brown, H. C. (2011). Gender, climate change and REDD+ in the Congo Basin forests of Central Africa. *International Forestry Review*, *13*(2), 163–176. doi:10.1505/146554811797406651

Corfee-Morlot, J., Parks, P., Ogunleye, J., & Ayeni, F. (2022). *Achieving clean energy access in sub-Saharan Africa*. OECD. https://www.oecd.org/environment/cc/climate-futures/Achieving-clean-energy-access-Sub-Saharan-Africa.pdf

Driessen, P. (2009). The real climate change morality crisis, Energy & Environment, 20(5), 765.

Electricity production from renewable, excluding hydroelectric (% of total) - Sub-Saharan Africa | Data. (2022). World Bank. https://data.worldbank.org/indicator/EG.ELC.RNWX.ZS?locations=ZG

Electricity production from renewable sources, excluding hydroelectric (% of total) - Sub-Saharan Africa. (2022). World Bank. https://data.worldbank.org/indicator/EG.ELC.RNWX.ZS?locations=ZG

GDP growth (annual %) - Sub-Saharan Africa. (2022). World Bank. https://data.worldbank.org/indicator/NY.GDP.MKTP.KD.ZG?locations=ZG

Guillaumont, P., & Simonet, C. (2011). To What Extent Are African Countries Vulnerable to Climate Change? Lessons from a New Indicator of Physical Vulnerability to Climate Change. *Fondation pour les études et recherches sur le développement international, 5.* https://ferdi.fr/dl/df-rKATnzmJv2KH9SKi8eijFqK7/ferdi-i08-to-what-extent-are-african-countries-vulnerable-to-climate-change.pdf

Maizland, L. (2021). *Global Climate Agreements: Successes and Failures.* Council on Foreign Relations. www.cfr.org. https://www.cfr.org/backgrounder/paris-global-climate-change-agreements

Mallowah, S., & Oyier, C. (2022). The Environment and Climate Change Law Review. *The Law Reviews.* thelawreviews.co.uk. https://thelawreviews.co.uk/title/the-environment-and-climate-change-law-review/kenya#:~:text=i%20The%20Constitution%20of%20Kenya%202010&text=Article%2070%20reinforces%20the%20right,Land%20Court%20under%20Article%20162

Marigi, S. N. (2017). Climate Change Vulnerability and Impacts Analysis in Kenya. *American Journal of Climate Change, 6*(1), 54–72. doi:10.4236/ajcc.2017.61004

Michaelowa, A., Hoch, S., Honegger, M., & Friedmann, V. (2016). *TRANSITIONING FROM INDCs TO NDCs IN AFRICA.* African Development Bank. Retrieved from African Development Bank website. https://www.afdb.org/fileadmin/uploads/afdb/Documents/Publications/AfDB-CIF-Transitioning_fromINDCs_to_NDC-report-November2016.pdf

Net migration - Sub-Saharan Africa. (2022). World Bank. https://data.worldbank.org/indicator/SM.POP.NETM?locations=ZG

Nyoka, N. (2022). *Inga 3 hydroelectric scheme is a looming disaster.* New Frame. https://www.newframe.com/inga-3-hydroelectric-scheme-is-a-looming-disaster/

Oke, Y. (2021). Nigerian electricity law and practice, (2nd edition). Lagos, Nigeria, Princeton and Associates Publishing Co. Ltd.

Oke, Y. (2021). Nigerian energy resources law and practice, (2nd edition). Lagos, Nigeria, Princeton and Associates Publishing Co. Ltd.

Pereira, M. (2018). Energy poverty and climate change elements to debate. In: Debra J. Davidsons; Matthias Gross. (Org.). Handbook on Energy and Society. Oxford University Press.

Poverty headcount ratio at $1.90 daily (2011) PPP (% of the population) - Sub-Saharan Africa. (2022). World Bank. https://data.worldbank.org/indicator/SI.POV.DDAY?locations=ZG

Renewable energy consumption (% of total final energy consumption) - Sub-Saharan Africa. (2022). World Bank. https://data.worldbank.org/indicator/EG.FEC.RNEW.ZS?locations=ZG

Renewable energy consumption (% of total final energy consumption) - Sub-Saharan Africa. . (2022). World Bank. https://data.worldbank.org/indicator/EG.FEC.RNEW.ZS?locations=ZG

Rumble, O. (2018). Climate change legislative development on the African continent. *4th Scientific Conference of the Association of Environmental Law Lecturers from African Universities in cooperation with the Climate Policy and Energy Security Programme for Sub-Saharan Africa of the Konrad-Adenauer-Stiftung and UN Environment.* UN Environment.

REA. (n.d.). *Solar Power Naija - Enabling 5 Million new Connections.* REA. https://rea.gov.ng/solar-power-naija/#:~:text=The%205Million%20Solar%20Power%20Naija,II

Thiede, B., & Strube, J. (2020). Climate variability and child nutrition: Findings from sub-Saharan Africa. *Global Environmental Change*, *65*, 102192. doi:10.1016/j.gloenvcha.2020.102192 PMID:34789965

Total greenhouse gas emissions (kt of CO2 equivalent) - Sub-Saharan Africa. (2022). World Bank. https://data.worldbank.org/indicator/EN.ATM.GHGT.KT.CE?locations=ZG

UNECA. (2022). *Nationally Determined Contributions (NDCs).* United Nations Economic Commission for Africa. www.uneca.org. https://www.uneca.org/african-climate-policy-centre/nationally-determined-contributions-%28ndcs%29

United Nations. (2018). *According to UN experts, Indigenous rights must be respected during Kenya's climate change project.* OHCHR. https://www.ohchr.org/en/press-releases/2018/01/indigenous-rights-must-be-respected-during-kenya-climate-change-project-say

United Nations Framework Convention on Climate Change (UNFCCC). (2016). *The Paris Agreement.* UNFCC. https://unfccc.int/sites/default/files/resource/parisagreement_publication.pdf

USAID. (2021). *Nigeria, Climate Change Country Profile.* USAID. https://www.usaid.gov/climate/country-profiles/nigeria#:~:text=Its%20multiple%20ecological%20zones%20have,to%20flooding%20and%20waterborne%20disease

World Bank. (2020). *Doing Business.* World Bank. https://documents1.worldbank.org/curated/en/688761571934946384/pdf/Doing Business 2020-Comparing-Business-Regulation-in-190-Economies.pdf

World Bank. (2021). *Nigeria to Improve Electricity Access and Services to Citizens.* World Bank. https://www.worldbank.org/en/news/press release/2021/02/05/Nigeria-to-improve-electricity-access-and-services-to citizens

WorldBank. (2022). *Sub-Saharan Africa* (pp. 4–17). Washington, D.C., United States: WorldBank. https://databank.worldbank.org/data/download/poverty/33EF03BB-9722-4AE2-ABC7-AA2972D68AFE/Global_POVEQ_SSA.pdf

Yong, C. (2022). *About Africa.* UNDP in Africa. https://web.archive.org/web/20200411014537/https://www.africa.undp.org/content/rba/en/home/regioninfo.html https://climatepromise.undp.org/news-and-stories/without-respecting-rights-and-knowledge-indigenous-peoples-climate-pledges-will

Chapter 3
Integrated Resource Planning for Efficient Power Supply in Nigeria by 2050:
Future Scenario Approach

Olutosin A. Ogunleye
https://orcid.org/0000-0001-7058-8856
Nigerian Defence Academy, Nigeria

ABSTRACT

Power supply in Nigeria has been epileptic due to low capacity for electricity generation, transmission, and distribution. The country's peak dispatch power stands at 5,552.8 MW against estimated demand of about 98,000MW thus stifling socio-economic activities with attendant adverse effect on economic growth. The absence of effective integrated resource planning was identified as a major cause of the low level of power supply in Nigeria as available energy resources have not been optimally harnessed. Using a future scenario approach, this chapter developed plausible scenarios for Nigeria's power supply by the Year 2050 based on a predictive model of seven per cent (moderate growth), ten per cent (high growth) and 13 per cent (optimistic growth) used in analyzing the impact of energy growth on Nigeria's economy. This led to the generation of scenario-based strategies including the optimal diversification of electricity generation mix and restructuring of the power grid system in Nigeria.

DOI: 10.4018/978-1-7998-8638-9.ch003

INTRODUCTION

The efficient supply of power is a major stimulant of socio-economic activities for improved productivity and development of nations. The major productive sectors of a nation's economy depend on power supply from one or more sources of energy for efficient operation. The sources of power supply could be fossil fuels such as coal that facilitated mechanisation through steam engines which characterised the First Industrial Revolution of 1764-1840 (Onifade, 2020). In contemporary times, the major source of power supply is electricity which was developed during the Second Industrial Revolution (2IR) between 1870 and the early 1900s. In the 2IR era, electricity was mainly produced from fossil fuels such as coal and crude oil. In the Third Industrial Revolution (1969-1984), renewable sources of energy such as hydropower, solar and wind energies began to gain prominence. Contemporary power supply has also witnessed the incorporation of mini and micro grids to the conventional national grid systems. Consequently, Integrated Resource Planning (IRP) has been proven as a sustainable means of power supply for economic growth and development.

In Brazil, hydropower was the main source of power supply, providing about 8,930 Megawatts (MW) constituting 77 per cent of 11,600 MW of electricity to the national grid as at 2000 (Food and Agriculture Organisation, 2018). Low rainfall in 2001-2002 however led to power supply shortages thus impinging on the efficiency of power supply to industries with adverse effect on the country's economy (Jardini et al., 2012). Consequently, the government conducted the 2003-2004 energy sector reforms based on Integrated Energy Options (IEO) (Griebenow, 2019). By 2016, Brazil had built about 290 power plants from 4 energy sources: gas, hydro, solar and nuclear sources. Consequently, power supply increased by 191 per cent, from 21,202MW in 2008 to 61,708MW in 2020 (BNamericas, 2020). The improvement in power supply facilitated industrial growth, particularly steel production that rose from 13 million tons to 422.8 million tons, representing an increment of 3,152 per cent, between 2009 and 2019 respectively. As a result, the country's steel industry contributed about US$1.6 billion to the economy in 2019 (Fonseca, 2020). Thus, Integrated Resource Planning constituted a major enabler of improved industrialisation for economic growth in Brazil.

In 2010, South Africa formulated the Integrated Resource Plan (IRP) to diversify the electricity generation sources for its power supply system that was hitherto largely based on coal. Coal had constituted about 85 per cent of the country's 45,700 MW capacity for power supply as at 2014 (Craig, 2018). Consequently, the country began to integrate renewable and other non-renewable energy sources such as solar, wind and nuclear power into its power supply system (Bakar, 2015). This initiative facilitated 27 per cent increase in the country's electricity generation that rose from 2014 value to 58,095 MW in 2020 (USAID, 2021). The increment in electricity generation facilitated industrial sector growth, whose contribution to the country's Gross Domestic Product (GDP) increased from US$102.4 billion to US$106.8 billion between 2010 and 2019. This represented an increase of 3.5 per cent within the mentioned period (The World Bank, 2020). Thus, IRP boosted industrialisation and the economy of South Africa.

The Federal Government of Nigeria (FGN) initiated the National Integrated Power Project (NIPP) in 2004 to boost power supply in the country. The project comprised 7 gas power plants of about 4,000 MW capacity to complement the existing power plants in Nigeria (Awosope, 2020). The FGN also initiated the Sustainable Energy for All Action Agenda (SE4ALL-AA) in 2016 with Vision 30:30:30 aimed at generating 30,000 MW of electricity, out of which renewable energy would provide 9,000 MW constituting 30 per cent, by 2030 (Federal Ministry of Power [FMP], 2016). Despite the efforts of the FGN, power supply in the country has been inadequate as peak power supply is about 5,552.8 MW against estimated demand of about 98,000MW in 2020 as highlighted in Figure 1 (Transmission Company of Nigeria, 2021; Eleanya, 2021). This is largely attributable to lean electricity generation mix, as natural gas and hydropower together constitute about 99 per cent of the installed electricity generation capacity of about 13,000 MW. However, frequent disruptions in gas supply, periodic low water levels and frequent grid collapse due to weak infrastructure and over centralised grid hampers the efficiency of the power supply system (Obiora et al., 2019). Consequently, Nigeria has been unable to effectively power its industrial sector which requires about 33,000 MW of electricity (Adedeji, 2016). This indicates inefficient IRP whose current structure would be unable to adequately support the country's industrialisation agenda for a vibrant economy by 2050.

The main objective of this paper is to proffer strategies for Nigeria to effectively utilise IRP to boost power supply for economic growth by 2050. The paper employed the future scenario approach to project plausible scenarios and proffered scenario-based strategies for utilising IRP to boost power supply in Nigeria by 2050. The scenarios generated were based on a predictive model that projected impact of energy growth on Nigeria's economy as 7 per cent (moderate growth), 10 per cent (high growth) and 13 per cent (optimistic growth) postulated by Ogunmodimu (2012).

Figure 1. Estimated power demand versus supply in Nigeria
(Source: GIZ, 2015; TCN, 2020; Eleanya, 2021)

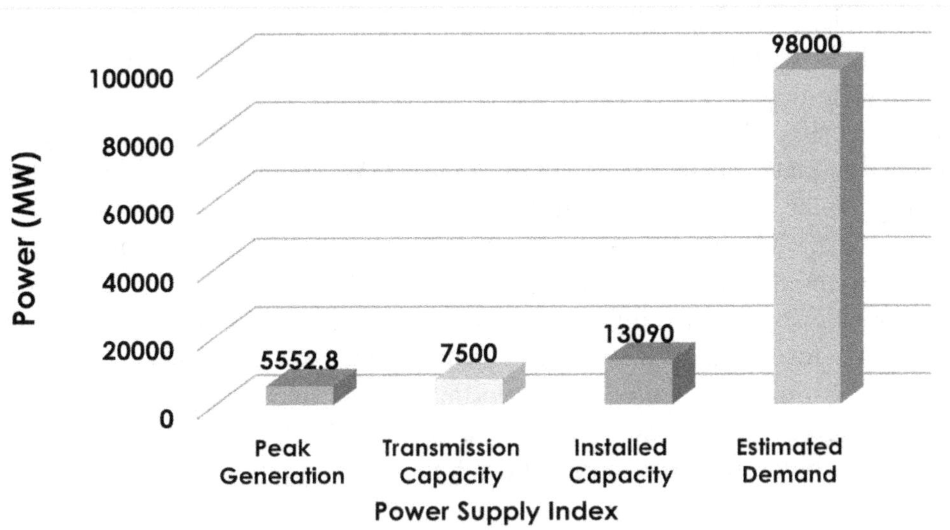

ENVIRONMENTAL SCAN

The scan of Nigeria's internal environment engaged the Base, Means and Capacity (BMC) analysis framework to derive the key decision factors while the external environmental scan utilised the Political, Economic, Social, Technology, Legal, Environment and Military (PESTLE-M) analysis.

Scan of the Internal Environment for Integrated Resource Planning Towards Reliable Power Supply in Nigeria by 2050

Nigeria's internal environment offers great potentials that could be exploited to optimise power supply for improved socio-economic development. These potentials are subsequently examined to derive key decision factors for IRP of the power supply system through BMC analysis.

Base

The Base analysis of Nigeria's internal environment covers the spatial, material and demographic bases.

Spatial Base

Nigeria lies between latitudes 4° and 14°N, and longitudes 2° and 15°E which falls within the global solar belt Its total area spans over 923,768 square kilometres (km2), comprising land area of 910,768 km2 and water body covering 13,000 km2 (CIA, 2020). The solar energy resource in Nigeria could adequately support about 427,000 MW of electricity generation, through solar PV and Concentrated Solar Power Technology (IIED, 2012; Craig, 2018). The country's coastline stretches over 853 km, providing wind speed of about 2 to 4 m/s at heights of 10m (Oyedepo, 2012). Wind speed of about 17.75m/s at a height of 50m is also obtainable in locations like Port Harcourt (Idris et al., 2020). These wind speeds fall within the minimum range of 3-5m/s and exceeds the optimal range of 10-15m/s required for generating electricity (Adaramola & Oyewola, 2011). This abundant wind resource would efficiently support offshore wind and tidal wave energies for a more robust power supply.

Material Base

Nigeria is favourably endowed with diverse energy resources, including fossil-based resources such as crude oil, natural gas and coal, as well as renewable energy resources such as solar, hydropower and wind. Its gas reserves amount to 187 trillion cubic feet (TCF), representing about 2.7 per cent of the 6,923 TCF of global gas reserves as at 2019 (Ogunlowo et. al, 2015). Despite its diverse energy resources, Nigeria's electricity generation mix is adjudged to be lean as only natural gas and hydropower accounts for about 99 per cent of its sources of power supply (Falobi, 2019). The prevalent poor diversification of energy resources for power supply has largely impinged on the efficiency of the country's power supply system.

Demographic Base

Nigeria has a population of about 218,541,212 people indicating a population density of about 236 persons per km2. Additionally, the country's population has been growing by about 2.5 per cent annually, increasing from 140.5 million to 213.4 million persons between 2005 and 2021. Furthermore, urban population increased from about 73.4 million to 104 million people from 2011-2019 (Figure 2), indicating about 41.7 per cent population growth (World bank, 2023). According to Aliyu and Amadu (2017), the trend of increasing urban population is attributable to relatively poorer infrastructure, particularly electricity, in rural areas. The Rural Electrification Agency (REA) supplies about 27 MW of power to rural areas, which is much lower than the estimated demand of 4,000MW in the rural areas (Oyedepo, 2012). This limits capacity for siting industries and other economy-stimulating infrastructures in rural areas. Thus, absence of robust IRP in Nigeria's electricity sector could impede the attainment of a vibrant economy in Nigeria by 2050.

Figure 2. Urban population increment in Nigeria, 2011-2019
(Source: World Bank, 2023)

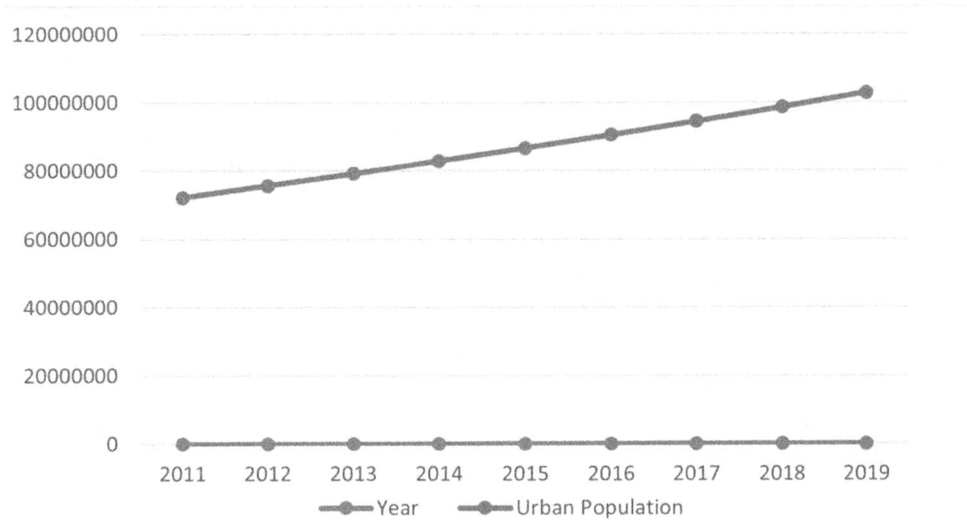

Means

The means consist of man-made competencies which covers economy, military, energy, political, technology, human capital and infrastructure.

Economy

Nigeria has the largest economy in Africa estimated at US$448.12 billion as at 2019 (World Bank, 2020). The economy is an emerging, middle-income and mixed economy broadly categorised into the agriculture, industry and services sectors (NBS, 2020). The Industrial Sector's contribution to the GDP fluctuated between 20.16 per cent and 27.38 per cent between 2012 - 2019. However, the manufacturing industry accounts for only 4 per cent of total industry contribution with the balance being contributed by the extractive industry, particularly Oil and Gas (Statista, 2020). This is largely attributable to inadequate power supply, which peaks at 5,552.8 MW against estimated demand of about 98,000 MW, thus unable to facilitate competitive production of goods. Therefore, failure to leverage IRP in Nigeria would likely limit manufacturing capacity thereby hindering industrialisation for a more robust economy by 2050.

Military

The Nigerian military, referred to as the Armed Forces of Nigeria (AFN), comprises 3 Services which are the Nigerian Army, Nigerian Navy and Nigerian Air Force. The combined strength of the AFN is about 215,000 personnel (GlobalFirePower, 2023). Some military cantonments, barracks or bases are in remote areas such as borders with neighbouring countries. Such areas usually have large expanse of land suitable for off-grid renewable power solutions such as solar and wind farms. These off-grid facilities could be sited close to military locations so that they would be protected from vandalism while also providing electricity to the surrounding community. Additionally, the military locations could be powered by off-grid solutions similar to the Energizing Education Programme that provides 28.5 MW of electricity to some tertiary institutions using solar hybrid and/or gas-fired captive power plants (Rural Electrification Agency, 2021).

Energy

Energy sources for power generation in Nigeria is largely based on fossil fuels, with natural gas accounting for about 78 per cent of the power generation mix (Oyedepo, 2012). The United Nations Sustainable Development Goal (UNSDG) 7 however advocates for clean energy for environmental preservation. However, Nigeria has not optimally exploited its renewable energy potential towards transiting from fossil fuels to renewable energy. For instance, the country only generated 28 MW and 3 MW of electricity from solar and wind energy respectively as at 2019 (IRENA, 2020). Thus, failure to implement robust energy transition from fossil-based fuels is an impediment to IRP for efficient power supply in Nigeria by 2050.

Political

In the political arena, the recent removal of power generation and transmission from the executive to the concurrent list is a commendable initiative which reflects the spirit of true federalism as enshrined in the Constitution of the Federal Republic of Nigeria (CFRN) 1999 (CFRN, 1999). This legal provision has addressed the hitherto arrangement which hindered State and local governments from embarking on power projects that could boost the country's IRP (Obiora et. al, 2019; FGN, 2023). Thus, leveraging of the new initiative on power generation and transmission by states and local governments offers good prospects for IRP towards enhancing the economy by 2050.

Technology and Infrastructure

Nigeria has not sufficiently developed indigenous technology to manufacture the infrastructure required in its power sector, leading to the importation of most of the required components and assemblies. It is essential to galvanise efforts of the research institutions in the country to domesticate the production of some of the components and assemblies utilised in the power sector through reverse engineering. Low hanging fruits in this regard are the production of solar photovoltaic cells and wind turbines whose production level are still relatively low. Government could also institute Research and Development programmes into emerging energy technologies such as Concentrated Solar Power (CSP) and offshore wind energy technologies that could also be exploited in Nigeria (Ogunmodimu & Okoroigwe, 2019). The low indigenous technological capacity in the power sector adversely affects electricity generation, transmission and distribution leading to inefficient power supply in the country. The existing transmission infrastructure has been unable to effectively support power transmission above 5,000 MW largely due to the aging infrastructure and over-centralisation of the transmission grid. This has led to partial or complete grid collapse occurring 564 times between 2000 and 2020 (Jimoh & Raji, 2023). Thus, overdependence on the aging and over-centralised electricity transmission grid would likely undermine efforts at leveraging IRP for efficient power supply in Nigeria by 2050.

Human Capital

The World Bank (2023) indicated that Nigeria's labour force was 73.27 million, representing 33.5 per cent of the country's population of about 218.54 million persons as of 2022. However, about 33.3 per cent of the workforce were unemployed as at First Quarter 2021 (NBS, 2021). An encouraging development is the exhibition of talent by some Nigerians in the development of indigenous prototype for electricity generation such as water-operated generators (The Cable, 2016). However, such efforts have not been adequately encouraged towards the development of home-grown power supply systems. Thus, promotion of indigenous R&D for local production of goods must be embraced for enhanced self-reliance in tackling perennial power supply issues by 2050.

Capacity

The capacity component of Nigeria's internal environment covers social, moral and political aspects.

Social

Nigeria is a very diverse socio-cultural society comprising over 250 ethnic groups with over 500 languages spoken in the country (CIA, 2020). Amongst these diverse ethnic groups, utilisation of wood as domestic source of energy is commonplace. This practise cuts across both urban and rural areas with over 70 per cent of the population depending on wood or charcoal for domestic energy supply (Adamu et al., 2020). Consequently, about 400,000 hectares of forest is lost annually in Nigeria due to exploitation of wood as an energy source. This reduces the capacity of the atmosphere to absorb greenhouse gases thus accelerating global warming and climate change (Adamu et al., 2020). Unfortunately, inadequate supply of electricity and other forms of clean energy at an affordable rate encourages the use of wood fuels which has adverse effects on human health and the environment. Thus, socio-economic development would be on the decline if IRP is not utilised to boost power supply in Nigeria by 2050.

Moral

Nigerians are generally intelligent, resilient and hardworking. However, the spirit of nationalism that engenders collective protection of public infrastructure is adjudged to be low (Dibua, 2017). This often leads to vandalism of power infrastructure, resulting into inefficient power supply partly attributed to sabotage of Government investments in the power sector. Security of power infrastructure is thus essential to grow the power sector towards appreciable economic growth by 2050.

Political

Bureaucratic processes often discourage potential investors in power projects in Nigeria (Ogunlusi, 2020). Failure to address issues of bureaucracy could constitute an obstacle to effective IRP for towards attaining robust economy by 2050.

Key Decision Factors

The scan of Nigeria's internal environment brought out some key decision factors such as renewable energy resources, energy transition, energy diversification, off-grid solutions, devolution of the power sector, emerging energy technologies, rural electrification, security and bureaucracy. The impact and uncertainty of these key decision factors are rated as highlighted in Table 1.

Table 1. Impact and uncertainty rating of the key decision factors

Key Decision Factors			Remarks
Factor	Impact	Uncertainty	
Renewable energy resources	High	Low	
Energy transition	High	High	
Energy diversification	High	High	
Off-grid solutions	High	Low	
Devolution of Power Sector	High	Low	
Emerging energy technologies	High	High	
Security	High	Low	
Bureaucracy	High	High	

Scan of the External Environment for Integrated Resource Planning towards Reliable Power Supply in Nigeria by 2050

The scan of Nigeria's external environment identified key driving forces that could impact on the efficiency of the country's power supply system for improved socio-economic development but are outside the sole purview of the country. These forces are subsequently examined through PESTEL-M analysis.

Political

The political scene from the context of the external environment was viewed from the lens of bilateral or multilateral arrangements in the power sector with other countries such as the West African Power Pool (WAPP). The WAPP is an arrangement for a common electricity market that would facilitate reliable power supply within the West African subregion. The arrangement encourages the development of all available power supply potential of participating countries whereby excess power could be sold through the WAPP. However, this arrangement is being hampered by inadequate capacity across the whole spectrum of the power sector amongst the member countries. For instance, the combined available power capacity in the 14 nations that comprises the WAPP was 12,666 MW as of Q4 2020 with Nigeria possessing 6705 MW representing about 52 per cent of this capacity (WAPP, 2023). The combined capacity of the 14 countries in the WAPP is less than the power supply requirement of Nigeria alone. This suggests that Nigeria must ramp up its power supply for domestic consumption with minimal reliance on external sources if its economy is to be vibrant by 2050. Sufficient capacity would also enable Nigeria to improve its capacity to export power through the WAPP.

Economy

Nigeria is the largest economy in Africa with a GDP of US$477.39 billion as of 2022 (The World Bank, 2023). The country's economy however largely depends on the petroleum sector which accounts for about 65 per cent of Government revenue and 88 per cent of foreign exchange earnings (KPMG, 2019). Although Nigeria is the largest economy in Africa, industries are relocating to neighbouring countries due to epileptic power supply and unfriendly business environment amongst others (Ohajianya et. al, 2014). Efforts of the FGN towards addressing these issues include the establishment of the Presidential Enabling Business Environment Council (PEBEC) and the Presidential Power Initiative (PPI). The PEBEC has facilitated improved ranking for Nigeria in World Bank's global Ease of Doing Business rating, from 146 to 131 position between 2019 and 2020 (KPMG, 2019). The PPI, which is a bilateral arrangement between Germany and Nigeria to be executed by Siemens, is designed to improve Nigeria's power supply from current 5,552.8 MW to 25,000MW by 2030 (The Africa Report, 2020). The successful implementation of the PPI is thus paramount to the attainment of efficient power supply towards enhancing Nigeria's economy by 2050.

Social

Nigeria has been a victim of social stigmatisation in the international arena largely due to its relatively high Corruption Perception Index (CPI). For instance, the country ranked between 139 and 149 out of 176 countries in the global CPI between 2014 and 2020 (Trading Economics, 2021). The investment climate in Nigeria is further aggravated by insecurity and scarcity of foreign exchange amongst others. This impinges on foreign direct investment for power projects with attendant adverse effects on reliable power supply in Nigeria (Imam, 2018). Thus, investment climate is a key factor in optimising Nigeria's capacity for IRP to effectively facilitate a vibrant economy by 2050.

Technology

Nigeria has a low indigenous technological base and depends largely on developed nations for its technology needs. This could be attributed to low level of R&D exemplified by Nigeria's ranking of 117 out of 131 countries featured in the Global Innovation Index 2020 (World Intellectual Property Organization, 2020). Thus, the country has largely been a consumer of foreign technologies, having invested about US$32 billion in foreign technologies between 2002 and 2017 (Shittu, 2018). Unfortunately, this huge investment has not facilitated technology transfer to Nigeria relative to countries like China and India. The indigenisation of foreign technologies in Nigeria's power sector is therefore essential for sustainable IRP towards facilitating a vibrant economy by 2050.

Legal

Nigeria is a signatory to the Paris Agreement on climate change that was signed in 2015 with 194 other countries. The Paris Agreement is a commitment by the signatories to apply measures that would limit the rise in mean global temperature to well below 2 °C above pre-industrial levels, and preferably limit the increase to 1.5 °C by reducing emissions by about 43 per cent by 2030 (UN, 2023). This necessitates energy transition from fossil fuels to renewable energy towards curbing the effects of climate change. Achieving the reduced emission target would require an effective regulatory framework with committed implementation arrangement to curb indiscriminate emission of greenhouse gases by local and international oil companies (IOC) in Nigeria (Ejiogu, 2013). This will require political will on the part of the Federal Government of Nigeria to implement existing legal instruments and policies on energy transition, particularly the Petroleum Industry Act 2021 to facilitate clean energy that would support a vibrant energy in Nigeria by 2050.

Environment

The country is prone to environmental pollution and degradation due to unwholesome practices by IOC and their local counterparts. This is further compounded by illegal logging that leads to deforestation. Most of the woods logged in Nigeria are exported to China, with 737.2 kilo hectares of tree cover, representing 7.34 per cent of total tree cover in Nigeria, removed between 2012 and 2020 (Global Forest Watch, 2021). This aggravates the impact of climate change such as drought and desertification resulting into reduced availability of water and threat to energy security (Pan et al., 2018). Thus, continuous deforestation hampers the attainment of sustainable energy security for a vibrant economy in Nigeria by 2050.

Military

Militaries globally are renowned for the development of technologies that later finds general application. Examples of such technologies include the internet, global positioning system and microwave heat amongst others (Chin, 2023). The military in Nigeria have institutions such as Nigerian Defence Academy and Air Force Institute of Technology where research could be undertaken on military technologies that are applicable to the energy sector. Such research could be undertaken through collaborations among the military, academia and industries either locally or overseas. Such research could focus on thermal power that is applicable in aircraft engines and electricity generation amongst others. Thus, strategic collaborations could boost Nigeria's IRP for efficient power supply by 2050.

Driving Forces

The scan of Nigeria's external environment brought out some key driving forces which are power supply roadmap, investment climate, PPI, technology transfer, energy security, climate change and strategic collaborations. The impact and uncertainty of these driving forces are rated as highlighted in Table 2. This sets the tone for the plausible scenarios which are discussed next.

Table 2. Impact and uncertainty rating of the key driving forces for Nigeria external environment

Driving Forces			Remarks
Force	**Impact**	**Uncertainty**	
Power Supply Roadmap	High	Low	
Investment Climate	High	High	
Presidential Power Initiative	High	High	
Energy Security	Low	Low	
Technology Transfer	Low	High	
Climate Change	High	High	
Strategic Collaborations	Low	High	

GENERATION OF PLAUSIBLE SCENARIOS FOR INTEGRATED RESOURCE PLANNING TOWARDS RELIABLE POWER SUPPLY IN NIGERIA BY 2050

The study undertook the analysis and ranking of the key decision factors and driving forces based on their degree of impact and uncertainty as highlighted in Table 3. An interplay of these factors and forces facilitated the realisation of the impact-uncertainty matrix indicated as Figure 3. Thereafter, the scenario-generation matrix was realised from the key decision factors and driving forces with the greatest influence on IRP, which were adjudged as energy diversification and investment climate. The scenario-generation matrix, highlighted in Figure 4 led to the development of the plausible scenarios.

Figure 3. Impact – Uncertainty matrix

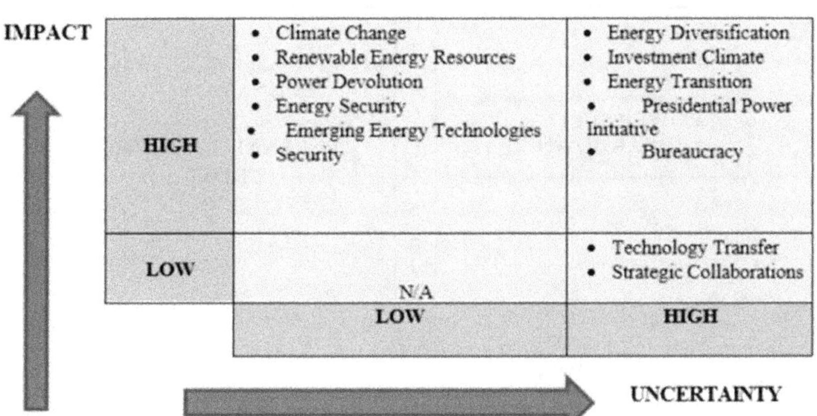

Table 3. Ranking table for key decision factors and driving forces

Key Decision Factors			Driving Forces		
Factor	**Impact**	**Uncertainty**	**Force**	**Impact**	**Uncertainty**
Renewable energy resources	High	Low	Investment Climate	High	High
Energy transition	High	High	Presidential Power Initiative	High	High
Energy diversification	High	High	Climate Change	High	Low
Off-grid solutions	High	Low	Power Supply Roadmap	High	Low
Devolution of Power Sector	High	Low	Energy Security	High	Low
Emerging energy technologies	High	High	Technology Transfer	Low	High
Security	High	Low	Strategic Collaborations	Low	High
Bureaucracy	High	High			

Figure 4. Scenario generation matrix

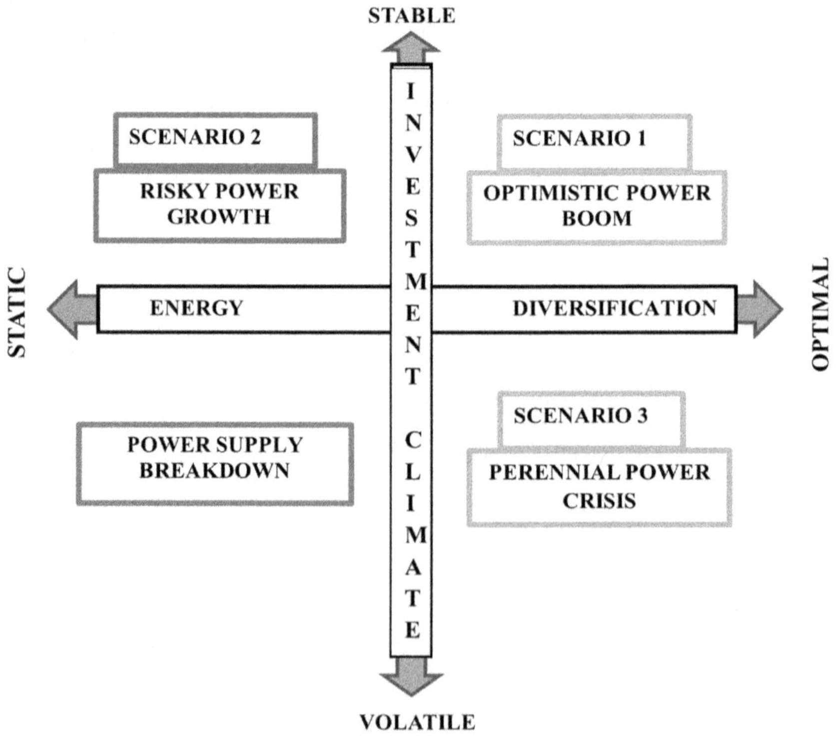

From Figure 4, the plausible scenarios generated are optimistic power boom, risky power growth and perennial power crisis. It is assumed that current and future Government efforts would prevent the fourth scenario, which is the total breakdown of the power supply system in the country.

Scenario One: Optimistic Power Boom

Scenario 1 describes optimal energy diversification in a stable investment climate. In this scenario, Nigeria leverages all available electricity generation resources including gas, hydro, nuclear, solar (PV and solar thermal), wind (on-shore, off-shore and tidal), biomass, hydrogen and geothermal. The national grid is also complemented by smart regional grids with independent but interconnected regional control stations as well as smart mini and micro grid systems in each of the 6 geopolitical zones. The smart decentralised transmission grids are connected to smart distribution systems which integrates net metering system that allows excess self-generated power from sources, such as solar home systems, to be fed back into the grid. Electricity market is highly liberalised to encourage private initiatives in major power projects and distributed generation through mini and micro grids. Electricity generation mix is highly diversified based on comparative advantage of available energy sources in different parts of the country. The scenario notes the existence of the PPI undertaken by Siemens Energy with the goal of ramping up transmissible power supply to 7,000 MW and 11,000 MW by 2021 and 2023 respectively and 25,000 MW by 2025. It however posits an extended timeline for the attainment of the targets taking cognisance of the fact that the current transmissible power hovers around 5,000 MW or even less. The scenario also notes the SE4ALL-AA of the Federal Government of Nigeria that highlighted modalities for attaining 45,101 MW power supply by 2030 (FMP, 2016).

The scenario projects that Nigeria would attain 7,000 MW of transmissible electricity based on the ongoing PPI by 2025 translating to electricity generation per capita of 0.0298kW based on projected population of 234.6 million persons by 2025. The World Bank (2023) further projects Nigeria's population between 2030 and 2050 to rise from about 262.6 million to 377.5 million as indicated at Table 4. Using the 13 per cent optimistic projection for energy growth compounded over the working timeframe and the base year as 2025, a model for deriving projected electricity generation in this scenario is highlighted in Equation 1. The projected electricity generation by 2050, under this optimistic scenario, is 237,799 MW as highlighted in Table 5.

$$E=P[x(1+r/n)^{nt}] \quad (1)$$

where E = Projected electricity generation
P = Projected population in reference year
x = Electricity generation per capita of base year (kW)
r = Projected growth rate
n = Number of times interest is compounded per year
t = Timeframe (years)

Table 4. Nigeria's projected population (2025-2050)

Year	Projected Population
2025	234,573,603
2030	262,580,426
2035	291,591,598
2040	320,779,660
2045	349,603,702
2050	**377,459,883**

Source: World Bank, 2023.

Table 5. Projected electricity generation under optimistic power boom scenario

Year	Projected Electricity Generation (kW)
2030	13,129,021
2035	29,159,160
2040	60,948,135
2045	118,865,258
2050	**237,799,726**

In this scenario, the projected massive growth in electricity generation is attributed to stable investment climate occasioned by renewed investors' confidence in the Nigerian economy, transfer of power sector matters from exclusive to concurrent list and removal of bureaucracies among others. The Relative higher cost of renewable energy could constitute an impediment to the realisation of the projection in this scenario. The plausibility of the scenario occurring is however enhanced by renewed Government focus and drive including the provision of incentives, including feed-in-tariff and power purchase agreements for renewable energy projects as highlighted in the

National Renewable Energy Action Plan (NREAP) 2015-2030. A comparison of the 237.8 GW projected electricity generation under this scenario with the 258 GW projection in the Nigeria Energy Transition Plan 2060, highlighted in Figure 5, indicates close values which suggests viability of the scenario.

In determining the capital cost of electricity generation by 2050 under this scenario, the percentage allocation of energy resource indicated at Table 6 was used as a guide in allocating percentages/weights to the energy resources considered under this scenario as indicated at Table 7. This facilitated the determination of projected capital cost for the various energy resource considered under this scenario. From Table 7, the minimum projected capital cost is $426.753 billion while the maximum is $853.529 billion. The minimum projected cost was computed with some incentive while the maximum capital cost did not consider any incentive (IRENA, 2012; Lazard, 2023). It was however noted that the minimum projected capital cost is closer to the €315.893 billion ($347.98 billion) computed using data from ECONSTOR than the maximum value. Furthermore, the minimum projected capital cost and the capital cost obtained from ECONSTOR data are closer to the projected capital cost of $270 billion power sector generation capacity and $135 billion for transmission and distribution infrastructure, a total of $405 billion proposed in the NTEP (NETP, 2022).

Figure 5. Comparison of optimistic scenario with NETP projection for electricity generation

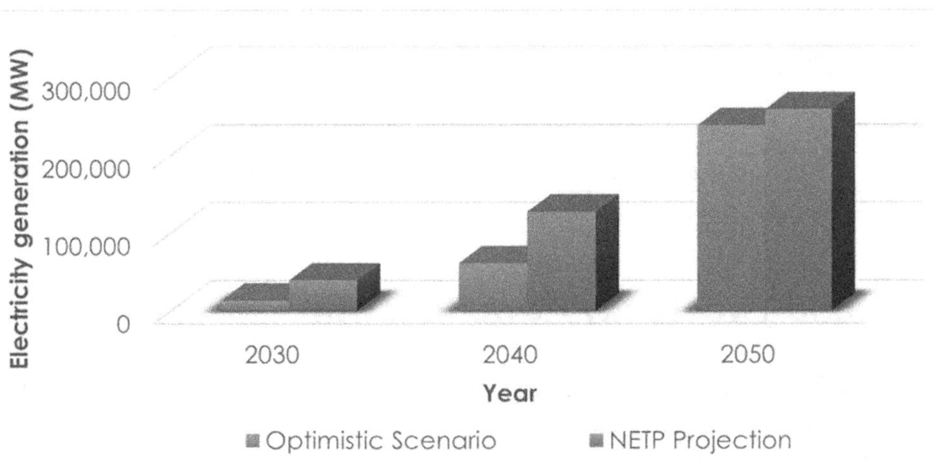

Table 6. Potential electricity generation mix by 2050 in NETP 2060

Year	Share of projection (GW)	% of projection
Hydro	11	4.26
Gas	10	3.88
Solar	197	76.36
Biomass	6	2.32
Hydrogen	34	13.18
Total	**258**	**100**

Source: NETP, 2022

Table 7. Projected capital cost for optimistic growth scenario by 2050

Year	% of projection	Share of projection (GW)	Unit capital cost-min ($/KW)	Unit capital cost-max ($/KW)	Unit capital cost (ECONSTOR) - (€/kW)	Projected capital cost -min (Billion $)	Projected capital cost - max (Billion $)	Projected capital cost (ECONSTOR) (Billion €)
Hydro	10.26	24.42	1050	7650	2000	25.64	186.813	48.84
Gas (CC)	20.88	49.69	650	1300	800	32.299	64.597	39.752
Solar PV	34.36	81.78	700	1400	425	57.25	114.492	34.757
Solar thermal (CSP)	3	7.14	7100	9800	1520	50.694	69.972	10.853
Wind (onshore)	2	4.76	1025	1700	1075	4.879	8.092	5.117
Nuclear	10	23.8	8475	13925	6000	201.705	331.415	142.8
Geothermal	2	4.76	4700	6075	2740	22.372	28.917	13.042
Wind (offshore)	2	4.76	3000	5000	2093	14.28	23.8	9.963
Biomass	2.32	5.52	1920	2440	1951	4.454	5.661	10.769
Hydrogen	13.18	31.37	1000	1500	-	13.18	19.77	-
Total	**100**	**238**				**426.753**	**853.529**	**315.893**

Sources: IRENA, 2012; Schröder, 2013; Lazard, 2023

Scenario Two: Risky Power Growth

Scenario 2 foresees poor energy diversification in a stable investment climate. This scenario is premised on failure to diversify the electricity generation mix from gas and hydro power plants, that jointly comprises about 99 per cent of the total electricity generation mix. Gas thermal plants alone accounts for about 83 per cent of installed power supply capacity in Nigeria (Falobi,

2019). Since the projected investment climate is stable, private investments in power projects would likely be undertaken but such projects would likely be limited to gas-fired thermal projects that are relatively cheaper than renewable sources such as solar and wind energy as highlighted in Table 8, although the indicated values of LCOE could vary based on factors such as location, capacity factor and incentives. Additionally, stable investment climate would likely facilitate the realisation of increased investments in conventional power plants, embedded generation and distributed generation that leverages enhanced gas distribution infrastructure. Relaxation of government monopoly in operating the transmission network, through the Transmission Company of Nigeria (TCN), would likely result into increased emergence of mini and micro grids including off-grid solutions. These off-grid solutions would likely involve solar PV and wind power though at relatively small capacities. Furthermore, direct power sale to customers by generation companies in line with the "willing buyer willing seller" policy of government would likely become more pronounced especially where generation companies are unable to feed the central grid with their optimal generation capacity. It is also envisaged that gas distribution infrastructure would be enhanced in view of the designation of gas as Nigeria's energy transition fuel.

This scenario is based on projected electricity generation growth of 10 per cent (high growth) from the base year of 2025. It also envisages that the PPI would deliver 7,000 MW dispatchable power by 2025. From Equation 1, the projected electricity generation by 2050 under this scenario is 120,787 MW among others as highlighted in Table 9. The scenario could be hampered by insecurity issues such as vandalization of gas supply infrastructure and low water levels for hydropower occasioned by climate change. Failure of the PPI and insecurity could also relegate the projected outcome of this scenario to the current epileptic power supply prevalent in the country. The minimum projected capital cost obtained for this scenario is $82.857 billion while the maximum and the capital cost obtained from Schröder (2013) are $271.432 billion and $117.693 billion respectively (Table 10). It should however be noted that the lower capital cost obtained in this scenario relative to optimistic growth scenario is due to the lower electricity generation capacity in this scenario.

Table 8. Levelized cost of electricity for selected energy sources

Electricity Source	Levelized Cost of Electricity (US$/kW) By Different Agencies		
	NERC	IEA	US DoE
Open Cycle Gas Turbine	0.065	0.048	0.05
Combined Cycle Gas Turbine	0.062	0.05	0.048
Solar Photovoltaic (PV)	0.13	0.17	0.18
Solar Thermal	0.27	0.19	0.21
Biomass	0.08	0.10	0.14
Onshore Wind	0.08	0.14	0.09
Nuclear	0.09	0.12	0.08
Large Hydro	0.04	0.07	0.05

Table 9. Projected electricity generation under risky power growth scenario

Year	Projected Electricity Generation (kW)
2030	13,129,021
2035	23,327,328
2040	38,493,559
2045	69,920,740
2050	**120,787,163**

Table 10. Projected capital cost for risky power growth scenario by 2050

Year	% of projection	Share of projection (GW)	Unit capital cost-min ($/KW)	Unit capital cost-max ($/KW)	Unit capital cost (ECONSTOR) - (€/kW)	Projected capital cost -min (Billion $)	Projected capital cost - max (Billion $)	Projected capital cost (ECONSTOR) (Billion €)
Hydro	15	18.12	1050	7650	2000	19.026	138.618	36.24
Gas (CC)	80	96.64	650	1300	800	62.816	125.632	77.312
Solar PV	3	3.62	700	1400	425	2.534	5.068	1.539
Wind (onshore)	2	2.42	1025	1700	1075	2.481	4.114	2.602
Total	**100**	**120.8**				**86.857**	**271.432**	**117.693**

Scenario Three: Perennial Power Crisis

Scenario 3 connotes perennial power crisis in which Nigeria's power supply by 2050 cannot adequately support the population and economy. Although there are policies and action plans to diversify the sources of electricity generation in the country, poor investment climate due to corruption and bureaucracies would likely discourage potential investors in the power sector. Renewable energy sources are somewhat exploited but poor implementation of relevant government policies and plans due to bureaucracies and corruption constitutes an impediment to this scenario. The poor investment climate also hampers the implementation of the PPI thereby limiting dispatchable power supply to 5,500 MW by 2025, resulting in electricity generation per capita of 0.023kW by 2025. Tenacity of investors could however facilitate the attainment of 7 per cent growth (moderate growth) in the power supply system between 2025 and 2050, leading to total available power supply of about 27,127 MW by 2050. The potential energy sources under this scenario are gas, hydro, solar PV and biomass. The projected power supply of about 27 GW would be abysmally low for Nigeria's projected population of about 377 million people by 2050 thereby impinging on the growth capacity of the nation's economy by 2050.

The plausibility of the scenario occurring is based on the country's poor corruption perception index where it ranked 150 out of 180 countries as of 2022 (Transparency International, 2023). This poor rating is despite government efforts in encouraging investments such as the establishment of the Presidential Enabling Business Environment Council (PEBEC) popular called the "ease of doing business" and the recent policy on bridging the gap in the foreign exchange regime. However, this scenario offers prospects of moving into Scenario 1 with the placement of effective measures to remove impediments to a favourable investment climate in the power sector. From Equation 1, the projected electricity generation by 2050 under this scenario is 20,787 MW among others as highlighted in Table 11. The minimum projected capital cost obtained for this scenario is $42.509 billion while the maximum and the capital cost obtained from ECONSTOR data are $220.161 billion and $67.046 billion respectively (Table 12). The relative lower capital cost obtained in this scenario is also due to the lower electricity generation capacity in this scenario.

Table 11. Projected electricity generation under perennial power crisis scenario

Year	Projected Electricity Generation (kW)
2030	7,877,412
2035	11,663,664
2040	19,246,779
2045	27,968,296
2050	**45,295,185**

Table 12. Projected capital cost for perennial power crisis scenario by 2050

Year	% of projection	Share of projection (GW)	Unit capital cost-min ($/KW)	Unit capital cost-max ($/KW)	Unit capital cost (ECONSTOR) - (€/kW)	Projected capital cost -min (Billion $)	Projected capital cost - max (Billion $)	Projected capital cost (ECONSTOR) (Billion €)
Hydro	55	24.92	1050	7650	2000	26.166	190.638	49.84
Gas (CC)	30	13.59	650	1300	800	8.833	17.667	10.872
Solar PV	10	4.53	700	1400	425	3.171	6.342	1.925
Biomass	5	2.26	1920	2440	1951	4.339	5.514	4.409
Total	**100**	**45.3**				**42.509**	**220.161**	**67.046**

SCENARIO BASED STRATEGIES FOR LEVERAGING INTEGRATED RESOURCE PLANNING TO BOOST POWER SUPPLY IN NIGERIA BY 2050

The scenario-based strategies to for effective IRP to boost power supply in Nigeria by 2050 include the implementation of extant policies and programmes on enhancing power supply, optimal diversification of electricity generation mix and restructuring of the power grid system in Nigeria. These strategies are discussed subsequently.

Implementation of Extant Policies and Programmes on Enhancing Power Supply in Nigeria

The realisation of the optimistic power boom scenario is hinged on the implementation of existing policies and programmes on boosting power supply in Nigeria. These policies and programmes include the Nigeria Energy Transition Plan, SE4ALL-AA, PPI and the National Renewable Energy

Action Plan 2015-2030 amongst others. A key factor to the realisation of this scenario is the execution of the Nigeria Energy Transition Plan where the country intends to attain about 258,000 MW of power generation with 197 GW of this figure obtained from solar, biomass and hydropower by 2050, although the terminal date of the plan is 2060 (Energy Transition Office, 2022). This initiative requires 10 billion NGN annually translating to 300 billion NGN between 2021-2050. A $23 billion investment opportunity has already been identified across a portfolio of projects/programmes in this regard. The portfolio reflects opportunities identified across the value chain, including generation, transmission and distribution, metering and gas commercialisation, amongst others. The funding for the optimistic scenario is thus dependent on significant implementation of the NETP 2060 while not neglecting other government initiatives to boost power supply in the country.

The optimal exploitation of solar energy for electricity generation with potential of about 427,000 MW and the completion of the 614 km Ajaokuta-Kaduna-Kano Gas Pipeline to facilitate the development of gas-fired thermal plants in the central and northern parts of Nigeria are also pertinent for the realisation of the NETP. Furthermore, the realisation of this scenario is dependent upon capacity to transmit all of the 11,154 MW installed capacity of thermal power plants and 1,936 MW installed capacity of hydro power plants in the next 5-7 years. It would also require the completion/execution of the proposed/ongoing power projects such as the 1,100 MW solar electricity projects, 3,050 MW Mambilla hydropower and 4,100 MW power purchase agreements initiated by the Nigeria Bulk Energy Trader (NBET) amongst others (FMP, 2020). Utilisation of nuclear energy for electricity generation would also boost the attainment of projections under the optimistic scenario.

Optimal Diversification Of Electricity Generation Mix

The impediments occasioned by poor energy diversification in the risky power growth scenario requires optimal harnessing of available power sources in the country. Apart from common energy sources such as gas, hydro, solar PV and onshore wind, opportunities abound for electricity generation through other energy sources such as concentrated solar thermal, geothermal, offshore wind, tidal and nuclear energy. Although studies have indicated that 427,000 MW of electricity is realisable from solar energy in Nigeria while hydropower could provide over 14,000 MW, the country only realises about 73 MW from solar PV and 1225 MW from hydropower (FMP, 2022; IRENA, 2023). Opportunity thus abounds for electricity generation from solar PV and hydropower, and this could be optimised through the implementation of the NETP. Similarly,

Nigeria needs to prioritise the development of nuclear energy for electricity generation purposes starting with deliberate steps to implement the proposed 4,800 MW nuclear power plants by 2035 (Mrabure, 2022). Ogunleye (2021) also determined the feasibility of utilising concentrated solar power of 50 MW -100 MW capacity at various sites in the northeast and northwest geopolitical zones of Nigeria. Additionally, Ohunakin et al (2023) posited the viability of offshore wind energy for electricity generations at selected sites in the Gulf of Guinea within Nigeria's territory. The wind resources obtainable offshore could also be utilised to produce renewable hydrogen. Diversifying electricity generation mix could be achieved under this scenario by leveraging the sources of energy highlighted in the optimistic growth scenario. Furthermore, the FMP could leverage the comparative advantage of each geopolitical zone in terms of renewable energy resource for the IRP model, as highlighted in Table 13. The proposed IRP model could be funded through a Build-Own-Operate PPP arrangement leveraging the Sovereign Green Bond domiciled at the Federal Ministry of Environment amongst other financing instruments. These measures would likely mitigate the risks associated with the risky power growth scenario towards a reliable power supply in Nigeria by 2050.

Table 13. Suggested renewable energy projects for geopolitical zones in Nigeria

Geopolitical Zone	Suggested Renewable Energy Project	Remarks
South West	Solar PV, Off-Shore Wind, Biomass, Small Hydro	Off-Shore Wind for coastal areas only
South East	Solar PV, Biomass, Small Hydro	
South South	Solar PV, Off-Shore Wind, Biomass, Small Hydro	
North West	Solar PV, On-Shore Wind, Biomass, Solar Thermal	
North East	Solar PV, On-Shore Wind, Biomass, Solar Thermal Large Hydropower	Large Hydro (Proposed Mambilla Hydropower)
North Central	Solar PV, On-Shore Wind, Biomass, Solar Thermal, Large Hydro	

Restructuring of the Power Grid

The study identified the need to restructure the power grid system in Nigeria to cope with the very significant improvement in load capacity envisaged by 2050. The objective of this proposal is to emplace efficient as well as smart electricity transmission and distribution systems to complement the existing grid. The envisaged efficient power grid system in 2050 would require additional control stations at sub-national levels. The restructuring of the grid would require the upgrade of the current national grid to a smart grid for optimal power utilisation. The proposed smart grid could also incorporate smart distribution systems capable of employing smart metering to encourage self-generated energy that could be fed back into the distribution system. Additionally, the proposed smart grid systems could commence with a pilot project in one city in each of the 6 geopolitical zones. Furthermore, privatisation of the Transmission Company of Nigeria to complete the unbundling process of the power sector from public to private management would offer good prospect for efficient power supply. This is to facilitate improved efficiency in power supply in view of the history of poor management of utilities by the public sector. The Federal Executive Council (FEC) could direct the Infrastructure Concession Regulatory Commission (ICRC) and the Federal Ministry of Power to work out modalities for the implementation of the proposed restructuring of the grid system. The services of a notable consultancy firm could also be considered in this regard.

CONCLUSION

The paper examined IRP as an imperative for a robust Nigerian economy by 2050 with a view to proffering strategic options for the nation. The environmental scan of Nigeria's internal environment using the BMC analysis revealed that Nigeria is richly endowed with both fossil and renewable energy resources but has not effectively harnessed these resources for efficient power supply. The sub-optimal level of power supply in Nigeria from the internal scan highlighted energy diversification as the key decision factor. The scan of the external environment using the PESTLE-M Analysis tool identified investment climate as the key driving force for efficient power supply in Nigeria by 2050. The interplay of the key decision factor and driving force produced 3 plausible scenarios regarding power supply in Nigeria by 2050. These scenarios are optimistic power boom, risky power growth and perennial power crisis. These plausible scenarios facilitated the generation of some scenario-based strategies which are the implementation of extant policies and programmes on enhancing power supply, optimal diversification of electricity generation mix and restructuring of the power grid system in Nigeria.

REFERENCES

Abam, F. I., Nwankwojike, B. N., Ohunakin, O. S., & Ojomu, S. A. (2014). Energy resource structure and on-going sustainable development policy in Nigeria: A review. *International Journal of Energy and Environmental Engineering*, *5*(1), 91–102. doi:10.100740095-014-0102-8

Adamu, M. B., Adamu, H., Ade, S. M., & Akeh, G. I. (2016). Household Energy Consumption in Nigeria: A Review on the Applicability of the Energy Ladder Mode. *Journal of Applied Science & Environmental Management*, *24*(2), 237–244. doi:10.4314/jasem.v24i2.7

Adaramola, M. S., & Oyewola, M. O. (2011). Wind speed distribution and characteristics in Nigeria. *ARPN Journal of Engineering and Applied Science*, *6*, 82–86.

Awosope, C. A. (2020). Nigerian Electricity Industry: Issues, Challenges and Solutions. *Public Lecture Series*, *3*(2), 2–14.

Bakar, L. (2015). *Governing Electricity in South Africa: Wind, Coal and Power Struggles*. University of East Anglia.

BNamericas. (2020). Brazil power generation back to growth. *BN Americas*. https://www.bnamericas.com/en/news/brazil-power-generation-back-to-growth

Central Intelligence Agency (CIA). (2020). Country profile: Nigeria. *CIA World Factbook*. CIA. https://www.cia.gov/library/publications/the-world-factbook/geos/sf.html

Chin, W. (2023). *War, Technology and the State*. Bristol University Press.

Craig, O. (2018). *Concentrating Solar Power (CSP) technology adoption in South Africa*. [PhD. Thesis. Faculty of Engineering. Stellenbosch University, South Africa].

DeCesaro, J. (2021). *Energy Transitions Initiative*. Office of Energy Efficiency & RE. https://www.energy.gov/eere/about-us/energy-transitions-initiative

Dyson, G. (2020). The South African Power Sector: Energy Insight. *Energy and Economic Growth*. https://energyeconomicgrowth.org/sites/eeg.opml.co.uk/

Eleanya, N. (2021). How to get rural households out of energy poverty in Nigeria: A contingent valuation. *Energy Policy*. https://www.sciencedirect.com/science/article/abs/pii/S0301421520307837

Falobi, E. O. (2019). The Role of Renewables in Nigeria's Energy Policy Mix. *IAEE Energy Forum*, (PP. 41-46). IEEE.

Falola, T. (2018). *A History of Nigeria*. Routledge.

Federal Government of Nigeria (FGN). (2023). *Electricity Act 2023*. FGN.

Federal Ministry of Power (FMP). (2016). *Sustainable Energy for All Action Agenda (SE4ALL-AA)*. Federal Ministry of Power.

FMP. (2022). *Ministerial Performance Report: Implementation of identified priorities and deliverables of Federal Ministry of Power*. FMP.

Fonseca, C. D. (2020). *Brazilian industrialisation: Notes on the historiographical debate*. Scielo. https://www.scielo.br/scielo.php?pid

Food and Agriculture Organisation. (2018). *Analysis and Systematization on Intended Nationally Determined Contributions in Latin America and Caribbean Countries*. FAO.

Global Fire Power. (2019). *Comparison Results: Military comparison results showcasing Nigeria and South Africa in side-by-side format*. Global Firepower. https://www.globalfirepower.com/countries-comparison-detail.asp?form=form&country1=nigeria&country2=south-africa&Submit=COMPARE

Griebenow, C., & Ohara, A. (2019). *Report on the Brazilian Power System*. Agora Energiewende. https://www.agoraenergiewende.de/fileadmin/projekte/2019/brazil_country_profile/155_countryprof_brazil_en_web.pdf

Habib, S. L., Idris, N. A., Ladan, M. J., & Mohammad, A. G. (2012). Unlocking Nigeria's Solar PV and CSP Potentials for Sustainable Electricity Development. *International Journal of Scientific and Engineering Research*, *3*(5), 1–8.

Hart, D. (2020). *Deployment of Solar Photovoltaic Generation Capacity in the United States*. US Department of Energy.https://www.energy.gov/sites/prod/files/f34/Deployment%

Imam, M. I. (2018). *Power Sector Reform and Corruption: Evidence from Sub-Saharan Africa*. EPRG. https://www.eprg.group.cam.ac.uk/wpcontent/uploads/2018/01/1801-Text.pdf

International Hydropower Association. (2022). *Hydropower Status Report*. IHA. https://www.hydropower.org/country-profiles/nigeria

International Institute for Environment and Development (IIED). *Renewable Energy Potential in Nigeria*. IIED. https://www.iied.org/sites/default/files/pdfs/migrate/G03512.pdf

IRENA. (2012). Renewable Energy Technologies: Cost Analysis Series. Abu Dhabi: International Renewable Energy Agency (IRENA).

IRENA. (2020). Renewable Capacity Statistics. Abu Dhabi: International Renewable Energy Agency (IRENA).

IRENA. (2023). Renewable Energy Statistics 2023. Abu Dhabi: International Renewable Energy Agency (IRENA).

Jardini, J. A., Ramos, D., Martini, J., Reis, L., & Tahan, C. (2012). Brazilian Energy Crisis. *Power Engineering Review*, *22*(2), 1–24.

Jimoh, M. A., & Raji, B. S. (2023). Electric Grid Reliability: An Assessment of the Nigerian Power System Failures, Causes and Mitigations. *Covenant Journal of Engineering Technology*, *7*(1), 17–20.

KPMG. (2019). *Nigeria ranks 131 in World Bank's 2020 Doing Business Report*. KPMG. https://kpmg.com/ng/en/home/insights/2019/10/nigeria-ranks-131-in-world-bank-s-2020-doing-business-report.html

Macrotrends (2020). Nigeria Literacy Rate 1991-2020. *MacroTrends*. https://www.macrotrends.net/countries/NGA/nigeria/literacy-rate

Mrabure, K. O. (2022). Nuclear Power as a Source of Energy in Nigeria and Sustainability of the Environment, NIALS. *Journal of Environmental Law*, *5*, 235–275.

Nick, M. (2020). *Energy and GDP in Nigeria*. Stanford. http://large.stanford.edu/courses/2016/ph240/nick2/

Nosiri, U. D., & Ohazurike, E. U. (2016). Border Security and National Security in Nigeria. *Southeast Journal of Political Science*, *2*(2), 214–226.

NTEP. (2022). *Nigeria's pathway to achieve carbon neutrality by 2060*. Nigeria Energy Transition Plan.

Nwankwo, O. C., & Njogo, B. O. (2013). The Effect of Electricity Supply on Industrial Production Within the Nigerian Economy (1970 – 2010). *Journal of Energy Technologies and Policy*, *3*(4), 34–42.

Obiora, C. A., Chiamogu, A. P., & Chiamogu, U. P. (2019). *Power Devolution and Electricity Transmission in Nigeria: A Study in Resources Mobilization for Economic Development*. Advances in Social Sciences Research Journal.

Ogundipe, A. A., & Akinyemi, O. (2013). Electricity Consumption and Economic Development in Nigeria. *Journal of Business Management and Applied Economics*, *2*(4), 1–13.

Ogunleye, O. A. (2021). *Techno-Economic Analysis of Concentrated Solar Power Projects in Northern Nigeria.* University of Ibadan Postgraduate College. https://pgsds.ictp.it/xmlui/handle/123456789/1258

Ogunlowo, O. O., Bristow, A. L., & Sohail, M. (2015). Developing compressed natural gas as an automotive fuel in Nigeria: Lessons from international markets. *Energy Policy*, *76*, 7–17. doi:10.1016/j.enpol.2014.10.025

Ogunmodimu, O. (2012). *Potential Contribution of Solar Thermal Power to Electricity Supply in Northern Nigeria* [Thesis, University of Cape Town].

Ogunmodimu, O., & Okoroigwe, E. C. (2019). Solar thermal electricity in Nigeria: Prospects and challenges. Energy Policy. *Renewable Energy*, *128*, 440–448.

Ohajianya, A., Abumere, O., Owate, I., & Osarolube, E. (2014). Erratic Power Supply in Nigeria: Causes and Solutions. *International Journal of Engineering Science Invention*, *3*, 51–55.

Ohunakin, O. S., Matthew, O. J., Adaramola, S., Atiba, O. E., Adelekan, D. S., Aluko, O. O., Henry, E. U., & Ezekiel, V. U. (2023). Techno-economic assessment of offshore wind energy potential at selected sites in the Gulf of Guinea. *Energy Conversion and Management*, *288*, 117–110. doi:10.1016/j.enconman.2023.117110

Onakoya, A. (2013). Energy Consumption and Nigerian Economic Growth: An Empirical Analysis. *European Scientific Journal*, *9*(4), 25–33.

Onifade, O. W. F. (2020). Artificial Intelligence and National Development. University of Ibadan.

Oseni, P. (2011). An analysis of the power sector performance in Nigeria. *Renewable & Sustainable Energy Reviews*, *15*(9), 4765–4774. doi:10.1016/j.rser.2011.07.075

Oyedele, T. (2021). *Economic and fiscal implications of Nigeria's rebased GDP*. PWC. https://www.pwc.com/ng/en/publications/gross-domestic-product-does-size-really-matter.html

Oyedepo, S. O. (2012). On Energy for Sustainable Development in Nigeria. *Renewable & Sustainable Energy Reviews*, *16*(9), 2583–2598. doi:10.1016/j.rser.2012.02.010

Oyewo, A., Aghahosseini, A., Bogdanov, D., & Breyer, C. (2018). Pathways to a fully sustainable electricity supply for Nigeria in the mid-term future. *Energy Conversion and Management*, *178*(1), 44–64. doi:10.1016/j.enconman.2018.10.036

Schröder, A., Kunz, F., Meiss, J., Mendelevitch, R., & von Hirschhausen, C. (2013). Current and prospective costs of electricity generation until 2050, DIW Data Documentation, No. 68. Deutsches Institut für Wirtschaftsforschung, DIW.

Statista. (2020). *Nigeria: Distribution of gross domestic product (GDP) across economic sectors from 2009 to 2019.* Statista. https://www.statista.com/statistics/382311/nigeria-gdp-distribution-across-economic-sectors/

The World Bank. (2020). *GDP growth (annual %) - South Africa.* World Bank. https://data.worldbank.org/indicator/NY.GDP.MKTP.KD.ZG?locations=ZA

The World Bank. (2023). *Population total – Nigeria.* The World Bank. https://data.worldbank.org/indicator/SP.POP.TOTL?end=2022&locations=NG&start=2022&view=bar

The World Bank. (2023). *GDP (Current US$) – Nigeria*. The World Bank. https://data.worldbank.org/indicator/NY.GDP.MKTP.CD?locations=NG

The World Bank. (2023). *Databank – Population estimates and projections*. The World Bank. https://databank.worldbank.org/source/population-estimates-and-projections

Trading Economics. (2020). *Nigeria - Economically Active Population in Agriculture.* Trading Economies. https://tradingeconomics.com/nigeria/economically-active-population-in-agriculture-number-wb-data.html

Trading Economics. (2020). *Nigeria GDP1960-2019 Data*. Trading Economics. https://tradingeconomics.com/nigeria/gdp

Trading Economics. (2020). *Nigeria Population*. Trading Economics. https://tradingeconomics.com/nigeria/population

Trading Economics. (2021). *Nigeria Corruption Index1996-2020 Data.* Trading Economics. https://tradingeconomics.com/nigeria/corruption-index

Transmission Company of Nigeria. (2021). *TCN Successfully Transmits Enhanced All-Time Peak.* TCN. https://tcn.org.ng/blog_post_sidebar104.php

United Nations. (2023). *The Paris Agreement.* UN. https://unfccc.int/process-and-meetings/the-paris-agreement

US DoE. (2017). Annual Energy Outlook. United States Department of Energy.

USAID. (2020). *South Africa Power Africa Fact Sheet.* USAID. https://www.usaid.gov/powerafrica/south-africa

WAPP. (2023). *West African Power Pool.* WAPP. https://www.ecowapp.org/

WIPO. (2020). *World Intellectual Property Organization (WIPO) Global Innovation Index 2020.* WIPO. https://www.wipo.int/edocs/pubdocs/en/wipo_pub_gii_2020/ng.pdf

Chapter 4
Inside São Tomé and Príncipe's Energy Transition:
An Analysis of Challenges and Progresses

João Simões
https://orcid.org/0000-0003-1437-5527
City University of Macau, China

ABSTRACT

This chapter examines the outlook for energy transitions in São Tomé and Príncipe, a small island developing state in Africa. It considers diverse dimensions, including the country's energy profile, national policies, institutions, and emerging challenges and opportunities. Additionally, this chapter discusses the role of international cooperation. The author poses the following research question: what are the challenges faced by São Tomé and Príncipe in the energy transitions? To answer this, the chapter draws on a range of sources, including scientific and grey literature, as well as official national development plans and reports. São Tomé and Príncipe's unique geographical location, political commitment to combating global warming, and membership in key global and regional organizations position it favorably for energy transitions. By underscoring the limitations and difficulties of national transitions in the context of São Tomé and Príncipe, this chapter aims to contribute to a broader understanding of energy transitions in small island developing states.

DOI: 10.4018/978-1-7998-8638-9.ch004

1. INTRODUCTION

Small island developing states (SIDS) are diverse countries facing specific developmental and environmental challenges. There is no universal definition for SIDS, and the number of eligible states depends on the classification criteria. According to the United Nations Office of the High Representative for the Least Developed Countries, Small Island Developing States and Landlocked Developing Countries (UN-OHRLLS)[1], SIDS are a distinct group of 37 United Nations (UN) Member States and 20 Non-UN Members/Associate Members that face "unique social, economic and environmental vulnerabilities," mainly located in the Caribbean, the Pacific, Atlantic, and Indian Oceans, and the South China Sea.

Since the establishment of UN-OHRLLS in 2001, it has supported, monitored, and further implemented the Barbados Programme of Action (BPoA), launched in 1994; The Mauritius Strategy, adopted in 2015; and the SIDS Accelerated Modalities of Action (SAMOA) Pathway (2014), to build synergies with other internationally agreed development goals, including the implementation of the 2030 Agenda for Sustainable Development and The Paris Agreement. Leandro et al. (2023) systematize the common characteristics of SIDS into five domains (geographical, political, economic, socio-cultural, and environmental) that help understand these countries and their particularities that motivated the UN-directed focus. The authors identify the principal geographical characteristics and their impact on transport dependence, also recognizing the recent independence from colonizing powers and the consequent political instability. This analysis aids in identifying the weaknesses in SIDS economies, which include lack of infrastructure, dependence on external financing, workforce outflow, high levels of socioeconomic inequality, climate change, limited access to drinking water, and land degradation.

The UN-OHRLLS emphasizes the critical role of South-South and triangular cooperation to overcome challenges and help SIDS accelerate the implementation of SDGs and the SAMOA Pathway. The body expects cooperation to "mobilize the human and financial resources, technology, innovation, knowledge, and expertise that SIDS now need." The UN-OHRLLS's report, Good Practices in South-South and Triangular Cooperation for Sustainable Development in SIDS,[2] describes various models of international cooperation, including "South-South and triangular practices," to support SIDS in accelerating progress toward sustainable development (UNOSSC, 2021).

The International Renewable Energy Agency's IRENA (2019)'s report, A New World - The Geopolitics of the Energy Transformation, foresees the impact of the energy transformations on SIDS as follows:

"Small Island Developing States (SIDS) will benefit most if they adopt renewable energy sources rather than fossil fuels. The import of fossil fuels now amounts to 8% of their GDP. Many SIDS are also extremely vulnerable to the effects of climate change. SIDS possess ample renewable energy sources, and renewable technologies can meet most of their domestic energy needs. The shift would cut import bills, promote sustainable development, and increase their resilience. International cooperation to support SIDS' renewable energy ambitions is growing substantially, and 13 SIDS have established 60-100% renewable electricity targets."

Contemporary definitions of energy security harmonize economic development with social and environmental concerns. Therefore, energy security requires "synergetic inter-sectorial visions," encompassing diverse elements ranging from "capital flows, flows of ideas and people, to education, technologies, taxation, and international cooperation." (Simões et al., 2023, p. 23) Considering the challenges outlined above, energy security is of great importance for SIDS because these countries often heavily depend on imported fuels, which results in both economic vulnerability and environmental issues. For these nations, there is a pressing need for adaptive and resilient strategies that harmonize energy security, energy equity, and environmental sustainability.

Beginning in 2014, the United Nation's Economic Commission for Africa launched a program through its African Climate Policy Center (ACPC) to assist African small island developing states in assessing their susceptibilities to adverse effects of climate change and developing response strategies that lessen exposure for their citizens. Cabo Verde, Comoros, Guinea Bissau, Mauritius, São Tomé and Principe, and the Seychelles all underwent scoping missions in order to: (1) Evaluate the country's needs for climate change adaptation and mitigation; (2) Determine the most effective strategies to address damage and increase national resilience to climate change; (3) Discuss the gap with key stakeholders in the government and other important groups; (4) Decide on the support modalities and collaboration structure.

Key findings from the six African SIDS show considerable vulnerability to climate change, both to fast-onset processes like sea level rise and catastrophic occurrences like cyclones. Over the course of the next century, these changes are expected to worsen, and all SIDS will need assistance in order to evaluate and deal with the resultant damage. Although they have small populations and are geographically isolated, their susceptibility is not uniform and they have different experiences with climate change as well as different capacities to deal with them (United Nations Economic Commission for Africa, n.d.).

Against this background, this chapter examines the outlook for energy transitions in São Tomé and Príncipe, officially the Democratic Republic of São Tomé and Príncipe (Portuguese: República Democrática de São Tomé e Príncipe) a small island developing state in Africa, located on the Equator in the Gulf of Guinea. It considers diverse dimensions, including energy profiles, national policies, institutions, and emerging challenges and opportunities.

2. SÃO TOMÉ AND PRÍNCIPE OVERVIEW

São Tomé and Príncipe is a small island developing state with a fragile economy and classified as a lower-middle-income country, making it one of the smallest economies in Africa. The population is vulnerable to exogenous shocks, with over two-thirds living below the poverty line and about a third living on less than 1.90 USD daily. The average person lives on less than 3.20 USD a day. The country faces typical small island state challenges, such as vulnerability to natural shocks and climate change effects. Limited human capital and scarce tradable resources compound the challenges of generating sustainable growth and reducing poverty rates (World Bank, n.d.).

Despite the difficulties, São Tomé and Príncipe has progressed in recent years. According to the Third National Report on the Millennium Development Goals (MDG), the country has achieved at least three of the eight MDGs: (1) Universal Primary Education, with a net enrolment rate of 98% in 2015 against 80% in 1990; (2) Reduction of infant mortality to 38 per 1000 live births in 2015 from 89 per 1000 live births in 1990; (3) Reduction of maternal mortality to 76 per 100,000 in 2015 from 151.3 per 100,000 live births in 2005. However, problems persist in the level of training young people receive, particularly in high school. According to the data collected in the National Sustainable Development Plan [PNDS] 2020-2024, the gross schooling rate is only 58% for boys and 71% for girls (República Democrática de São Tomé e Príncipe, 2019).

Although two inhabited islands constitute the country, the vast majority of the 217,164 (2022) population lives on São Tomé (210,427), 25% of whom live in the capital city. About 75% of the two islands' population live in urban areas (Santo António is the main town on Príncipe Island), and the remaining 25% live in dispersed rural areas (CIA, n.d.). This demography is a challenge to electricity distribution. On the one hand, the overpopulation of towns downgrades the quality and efficiency of the electrical grids. On the other hand, the rural population dispersion inhibits full national coverage.

The World FactBook (CIA, n.d.) reports that in 2019 only 71% of the population had access to electricity: 87% of urban and 25% of the rural areas were electrified. Generating capacity was 28,000 kW, and consumption was 78 million kWh. São Tomé and Príncipe does not import (or export) electricity: 89.5% is produced using fossil fuels and 10.5% by hydroelectricity (2020). No solar, geothermal, wind, nuclear, or sea-based energy sources exist. The grid efficiency is questionable because of an estimated transmission/distribution loss of 11.9 million kWh (2019). Energy consumption per capita is 11.636 million Btu/person (2019), and CO_2 emissions are 173,000 metric tonnes (2019).

The two-island archipelago, with an area of approximately 1,000 km^2 located in the Gulf of Guinea, with an approximate population of 210 thousand people, largely depends on plantation agriculture, especially cocoa and coffee, and fisheries. However, tourism has increased because of recent foreign investment (24 million USD in 2020). The state depends on foreign aid (approximately 51 million USD in 2010) for infrastructure investment. Recent oil exploration may change this productive framework and the country's economy.

The economy is not diversified and is sensitive to market prices for cocoa, the main export product. The country's trade deficit does not benefit its economic situation, although there have been improvements in recent years. Public administration is weak, which impacts efficiency and public policy. There is still a lack of energy, roads, seaports, airports, and water supply infrastructure to support economic growth.

The World Bank considers São Tomé and Príncipe population small, making it difficult to produce efficient goods and services on the scale needed to meet demand in local and export markets. Furthermore, the indivisibility of public goods production and the dispersed population lead to costly public services. The economy is driven primarily by public expenditure via foreign aid and government borrowing, agriculture, tourism, and foreign investment fueled by oil production expectations (World Bank, n.d.).

The Gulf of Guinea is an energy "choke point" for nations looking to diversify their supply sources. Approximately 70% of Africa's oil and natural gas exploitation is in this region. Africa produces 13% of world oil and 7% of natural gas. Nigeria is the continent's third-largest oil exporter after Gabon and Angola. 6.5% of oil consumed in the European Union (EU) comes from the Gulf of Guinea. Due to EU sanctions against Russia following the invasion of Ukraine, this percentage is expected to rise (Leandro et al., 2023).

São Tomé and Príncipe is not yet exploiting oil or gas and does not have refinery capability. The country imports 1,027 bbl/day (2015) of refined petroleum products and consumes 1,200 bbl/day (2019) (CIA, n.d.). Although the country is on the equator and has the potential for renewable energy production, fossil fuels supply most of its mobility and electricity needs. According to the African Union Energy Commission´s (n.d.) Energy Balance 2020, "the total primary energy supply was 170 ktoe. Biomass (firewood and charcoal) is used heavily for cooking purposes. There is no oil refinery. As a result, all petroleum products including jet fuel, gasoline and kerosene have to be imported. The fuel comes mostly from an Angolan supplier that has an effective monopoly. There are no indigenous sources of oil, coal, natural gas or hydropower. The share of electricity consumption was households 77%, commerce and public sector 23%."

The first exploratory drilling in EEZs was completed in July 2022 by a company established by British Shell and Portuguese Galp[3]. However, it is essential to point out that there are safety issues associated with oil exploration. There are environmental and piracy issues. On 3 November 2022, NATO sponsored a webinar dedicated to security in the Gulf of Guinea and stated that piracy and illegal commerce are energy concerns.[4] Ambassador Namina Negnm, an official of The African Union and Director of The African Migration Observatory, highlighted that pirates were hijacking tanker ships to ransom the crews and sell the crude oil. Recently the Nigerian authorities discovered an illegal oil pipeline that had been operating for at least eight years, smuggling stolen oil to black-market actors. The issue became so serious that the United Nations issued a UN Security Council Resolution (UNSC RES 2634) to fight illegal activities in the Gulf of Guinea.

3. STRATEGIC AND DEVELOPMENT PLANS

The National Plan for Sustainable Development of São Tomé and Príncipe [PNDS] (2020-2024) draws attention to the risks to the São Tomé economy arising from possible conflicts resulting from political instability in the region, in addition to the risks associated with climate change. "The strategic location, extended coastline, and exclusive economic zone expose the country to new threats, such as terrorism, drug and human trafficking, international crime, illegal immigration, maritime piracy, counterfeiting, and money laundering." (República Democrática de São Tomé e Príncipe, 2019)

PNDS also mentions that the energy sector has a weak production capacity and poor energy stability, aggravated by the operating conditions and chronic deficit in internal capacities to maintain and recover generators. The Government is determined to reverse this situation, laying the foundations for the "sustainable and clean electrification of the country" with hydro, wind, and solar sources. PNDS recommends elaborating a national energy sustainability program with its principal axes of intervention being: institutional strengthening, business environment improvement, reform of the energy market organizational structure, investment in strategic infrastructure, development of renewable energies, and promotion of energy efficiency.

Despite the progress in the population's access to electricity, some difficulties persist in energy infrastructure, a fundamental factor for sustainable economic growth. Energy access is one of the most critical development challenges in São Tomé and Príncipe. In addition, less than 5% of the population has access to "clean cooking." Renewable energy only contributes 40.75% of consumption.[5]

As the PNDS states, São Tomé and Príncipe seeks to "increase the energy supply and reduce costs, through the use of its own sources, to meet the needs of companies and connect 95% of the country's locations to the electricity grid." The country will achieve this objective by "increasing solar and gas production capacity to cover 50% of the country's needs with clean energy by 2030." In addition, it will be necessary to expand transmission and distribution capacities and increase energy efficiency.

Energy will be one of the relevant issues for the development of São Tomé and Príncipe. Indeed, as stated by the World Bank (2023) in a recent publication, for the current year 2023:

"The electricity sector is marked by high cost and unreliability, representing a binding constraint to economic development. While access to electricity is relatively high, estimated at 84% of the population, power outages are frequent, and service is unreliable. The sector is hampered by the high costs associated with its mostly thermal generation and service delivery challenges. The effects of unreliable energy are felt throughout the economy as it limits productivity. The results can, for example, be observed in a lack of cold storage and adequate processing facilities for the agriculture and fisheries sectors while hindering the development of tourism and digital connectivity. Furthermore, the sector's continued arrears accumulation is a major vulnerability for the country's public finances and external position. In 2022, the electricity and water company (EMAE) arrears reached 32.2% of GDP (about a third of total public debt). Transformation of the electricity sector would address an immediate constraint to private sector development and service provision while stabilizing the country's fiscal outlook. Such a transformation would include increasing generation, moving towards green energy sources over time, and addressing governance issues linked to EMAE."

4. THE REGIONAL CONTEXT AND INTERNATIONAL COOPERATION

The World Bank (2023) presents policy recommendations for economic growth and São Tomé and Príncipe's resilience, and points out the constraints to the country's development. It defines tourism as the central axis of the strategy. "Looking ahead, nature-based tourism is an anchor to support the economic transformation toward a blue economy." Therefore, it defines five action priorities, among which "priorities for increasing access to electricity and water" stands out:

1. Provide reliable and affordable universal access to electricity and clean water;
2. Change the electricity generation mix away from diesel toward renewable sources;
3. Increase private sector investment;
4. Implement sustainable electricity and water tariffs that support expanded access and affordability;
5. Improve governance and address the structural deficiencies of EMAE.

Intervening in pursuing this priority presents several challenges, particularly in the energy sector, since the electricity supply is expensive and unreliable. In addition, the poor management and low operational performance of EMAE undermine efforts to recover costs. However, as the aforementioned World Bank document highlights, this intervention will generate opportunities as essential steps are being taken to improve the reach and quality of service provision. In anticipation of the carbon transition, the authorities have adopted a least-cost power development plan (LCPDP) to maximize the renewable energy share and progressively reduce diesel-based generation. The government has also developed a Green Energy Acceleration Plan, which forecasts that, by 2030, 50% of generation will be from renewable sources. Furthermore, solutions to strengthen São Tomé and Príncipe's agencies for more robust policymaking are being implemented, especially to improve information on each sector's costs and consequent pricing adjustment to generate financial sustainability in each of them.

In response to many of these energy sector challenges, São Tomé and Príncipe is progressing with projects as indicated in National Action Plan for Renewable Energies (PANER) (Ministério das Infraestruturas e Recursos Naturais, 2021):

a) A mini-hydropower project support program implemented by the African Development Bank;
b) A program of economic reform and support to the electricity sector (ERPSSP-I) implemented by the African Development Bank;
c) The Institutional Support and Energy Transition Programme (ETISP) implemented by the African Development Bank;
d) A project to promote renewable energy and energy efficiency investments in the electricity sector of São Tomé and Príncipe (UNIDO Energy Project) implemented by UNIDO;
e) São Tomé and Príncipe Electricity Sector Recovery Project implemented by the World Bank;
f) A project to promote hydroelectric energy through an approach integrating land and forest management in São Tomé and Príncipe (ENERGY Project) implemented by UNDP/GEF;
g) The project proposal Building Institutional Capacity for a Renewable Energy and Energy Efficiency Investment Program for São Tomé and Príncipe has been sent to the GCF[6] with the support of UNIDO in August 2021.

With international organizations' financial and technical support, São Tomé and Príncipe has gained essential knowledge concerning the current positioning of the electricity sector, particularly renewable energies. Furthermore, the country has defined the objectives and strategies it will follow throughout the energy transition to increase non-fossil energy sources and efficiency gains, improve technical management, and increase the sector's integration with other activities linked to the territory and its use of land, forest, sea, and waterways. The economic reforms advocated aim to increase macroeconomic stability and reduce poverty.

Several international partner institutions support the projects highlighted in PANER, including the African Development Bank sponsoring support programs for mini-hydro projects, economic reform of the electricity sector,[7] institutional support, and the energy transition.[8] This support involves developing concrete actions and technical support.[9]

Cooperation with UN agencies and programs allows São Tomé and Príncipe to carry out other projects included in PANER, as mentioned above in points d, f, and g.

The World Bank is another critical international partner financing the energy sector to reach two concrete objectives: increasing renewable energy generation and improving electricity supply reliability. To achieve these goals, the Power Sector Recovery Project consists of 4 key components:[10]

1. "Support for electricity institutional reform and sector planning;"
2. "Strengthening of the operational performance and governance of EMAE;"
3. "Investing in enhanced reliability of electricity generation, transmission, and distribution;"
4. "Project implementation support. This component will finance project implementation support, including training for staff of the implementing agency, AFAP, on procurement and fiduciary duties. Technical training, in particular on O&M issues, will be provided to EMAE technical staff supervising project implementation."

Another aspect of sustainable development for São Tomé and Príncipe must focus on clean energy for cooking. A significant part of the population lacks access to sustainable cooking fuel, instead relying on biomass (firewood) and charcoal. The World Bank estimates that 72% of the population uses solid fuel for cooking, with firewood used by 45.6% of households, charcoal by 26.5%, petroleum by 25.5%, and liquefied petroleum gas (LPG) by 1.5%. The cooking fuels must become compatible with PANER's National Plan for Renewable Energies in São Tomé and Príncipe (Period 2021-2030/2050). (Ministério das Infraestruturas e Recursos Naturais, 2021)

Deforestation and erosion are the principal negative impacts caused by using biomass as the principal cooking fuel at a domestic and commercial level (small bakery and catering industries). In addition to firewood, the population also uses locally produced charcoal for cooking. Institutions estimate that almost 75% of the wood consumed is exploited illegally without regulation or supervision. Currently, São Tomé and Príncipe is a natural greenhouse gas

(GHG) sink thanks to its forest and plant resources capturing and fixing carbon. That offsets the population's carbon footprint. However, if emissions continue to rise and the government fails to control illegal deforestation, the country risks becoming a net GHG emitter. Therefore, PANER aims to implement measures promoting GHG emission reduction, reduce deforestation and promote sustainable agricultural production to maintain vegetation cover and its unique character as a GHG sink. As stated above, the country's principal issues are improving the energy sector, general economic efficiency, and soil and vegetation cover management. Therefore, PANER has defined its principal goals as follows:

1. Renewable Energy Targets in the Power Grid: 72% of energy from renewable sources (MW) by 2030, including 49% solar, 18% hydroelectric, and 5% biomass. These will be maintained until 2050.
2. Energy Efficiency Targets: a) reduce energy demand by 8.7% by 2030 and 12.9% by 2050; b) Change cooking fuel so that by 2050, 10% of the population use solid fuel, 87.5% liquid fuel, and 2.5% other sources; c) Introduce electric transport from 2040.

PANER also defines the following complementary goals:

1. Greenhouse Gas Targets: Reduce emissions by 27% before 2030;
2. Universal Energy Access targets: Ensure that 100% of the population has access to electricity and sustainable energy for cooking by 2030.

Several reports have helped to identify São Tomé and Príncipe's sustainable development and energy goals: São Tomé and Príncipe 2030 Transformation Agenda (2015), based on the UN 2030 Agenda, the Multisectoral Investment Plan (2017), which identified disaster risks (including drought), the Transition Strategy for the Blue Economy (2019), which includes most sectors associated with energy and sustainable development issues, and the National Plan for Sustainable Development (2019), which highlights the "need to reverse the current energy situation with the application of EE [energy efficiency] measures and the gradual increase in RE [renewable energy], with a view to the sustainable and clean electrification of the country (using hydro, solar and wind sources)."

However, it is vital to highlight the UN's objectives for São Tomé and Príncipe to integrate into other international organizations, for example into the Economic Community of Central African States (ECCAS).[11] Within the scope of Under the Global Network of Regional Sustainable Energy Centres (GN-SEC) program, "the United Nations Industrial Development Organisation (UNIDO) supports the Economic Community of Central African States (ECCAS) and its eleven Members States in the establishment of the Centre for Renewable Energy and Energy Efficiency for Central Africa (CEREEAC), which aims to accelerate the energy and climate transition by providing support 'from the region for the region'. (…) The process included the development of a baseline and needs assessment and was closely coordinated with the efforts of the International Renewable Energy Agency (IRENA) to design a Renewable Energy Roadmap for Central Africa."[12]

The first projects with the involvement of CEREEAC were with São Tomé and Príncipe and, as was previously stated, point to its capacity surrounding renewable energy and energy efficiency:

- GCF-UNIDO project "Building institutional capacity for a renewable energy and energy efficiency investment program for São Tomé and Príncipe."
- GEF-UNIDO project "Strategic program to promote renewable energy and energy efficiency investments in the electricity sector of São Tomé and Príncipe."[13]

The involvement of smaller countries such as São Tomé and Príncipe in organizations of this nature is an advantage as they are currently taking place in the context of the energy transition and the fight against climate change. The creation of CEREEAC within the regional association ECCAS represents an effort to create a single market for traditional products and services and boost sustainable energy and cleantech fundamental to São Tomé and Príncipe's development, by boosting the circular economy, economic diversity, and the reduction of dependence on exports of primary products (cocoa and crude oil). In situations like this, international cooperation allows each partner to align its objectives with the those of the organization. This promotes a more robust regional coordination and a stronger commitment to achieving the end results. "There is a need for economies of scale and speed. In this context, regional, south-south, and triangular cooperation can become an important accelerator," as referred to in the UNIDO Global Network of Regional Sustainable Energy Centers (GN-SEC) program,[14] in which São Tomé and Príncipe participates.

Cooperation can take the characteristics of aid between countries. An example of this type of cooperation is that which stems from the 27th meeting of the United Nations Climate Change Conference of Parties (COP27) in November 2022 in Sharm El Sheikh, in which "the countries agreed to set up a specific fund on loss and damage for vulnerable countries, including SIDS, to address their existential threats" (IRENA, 2023).

IRENA highlights "that over 90% of the solutions to achieve the 1.5°C pathway involve renewable energy through direct supply, electrification, energy efficiency (EE), green hydrogen and bioenergy that can be deployed rapidly and at scale."[15] As seen above, similar to most SIDS, this is the path that São Tomé and Príncipe is following, as the aforementioned IRENA report states: "SIDS have shown climate leadership in this regard, reflected in the 34 updated Nationally Determined Contributions (NDCs). Of these 34 NDCs, at least 18 have underscored stronger pledges to amplify energy targets in the power and end-use sectors, such as transport, heating, and cooling". It should also be noted that IRENA has supported this path, "the leading global intergovernmental agency for energy transformation that serves as the principal platform for international cooperation, supports countries in their energy transitions, and provides state of the art data and analyses on technology, innovation, policy, finance, and investment."

IRENA's support for the development and implementation of the Nationally Determined Contribution (NDC) in São Tomé and Príncipe was explicitly directed toward:

- Solar City Simulator for São Tomé;
- Assessing the deployment of renewable energy solutions for healthcare facilities;
- A cost-effectiveness analysis of renewable energy technology options.

In summary, the PNDS 2020-2024, taking into account the country's general objectives and, in particular, those that point to the need to accelerate accessibility and energy transition, bets "on the strengthening of diplomacy, international cooperation and regional and international integration of São Tomé and Principe." That is in line with the country's position over recent years. It is also essential for the country to deepen relations with ECCAS, the EU, Portugal, Brazil, NAFTA, the US, Angola, South Africa, China, and others, within the framework of the African Union's Agenda 2063 and pay attention to the African Continental Free Trade Agreement (AfCFTA). The PNDS 2020-2024 recommends that the country must be in a position to take advantage of the opportunities that the existing global networks provide in crucial areas such as tourism, the environment, the production and development of knowledge, technology, international trade, telecommunications, financial services, energy, and transport.

5. INSTITUTIONAL FRAMEWORKS FOR ENERGY TRANSITIONS AND SUSTAINABLE DEVELOPMENT

It should be noted that the current energy sector structure reflects the profound reforms it has undergone since 2014, with the entry into force of the Legal Regime of the Organization of the Electric Sector (RJSE), approved by Decree-Law no. 26/2014. The state and private entities are part of the energy sector organization in São Tomé and Príncipe. The Ministry of Public Works, Infrastructure, Natural Resources and Environment (MOPIRNA) defines the public policy for the sector and is responsible for defining objectives and strategies. The Directorate-General for Natural Resources and Energy (DGERNE) is part of the Ministry. It is operationally responsible for executing policy and implementing the strategic plans associated with the sector, such as National Action Plan for Renewable Energies (PANER) and National Action Plan for Energy Efficiency (PANEE). Table 1 contains the relevant information.

As seen in Table 1, the state also has support for its activity in project implementation from the World Bank and the Fiduciary Agency for Project Administration (AFAP). In addition, other coordinating bodies serve the state and electricity sector, in particular, the Coordination Committee for the Electricity Sector Transformation Program (CC-PTSE), Technical Support Group for the Electricity Sector Transformation Program (GT-PTSE), and the National Sustainable Energy Platform (PNES). Finally, the state intervenes in the electricity sector through EMAE, a public company present at all stages, from generation to electricity distribution, as a vertical monopoly.

The private sector has many weaknesses and faces various difficulties. Through the establishment of associations, local actors seek to represent and promote the private sector. As stated in the ALER´s (2020) report Renewable Energy and Energy Efficiency in São Tomé and Príncipe: "with regard to the renewable energy sector, although it is possible to identify some demand from the private sector in terms of renewable energy solutions for the most energy-intensive companies and sectors, in terms of supply there

Table 1. Organization of the electricity sector in São Tomé and Príncipe

Entity	Role
Ministry of Public Works, Infrastructure, Natural Resources and Environment (MOPIRNA)	Define state policy for the sector Plan and manage the national electricity system
Directorate-General for Natural Resources and Energy (DGRNE)	Natural resources and energy policy sectors: elaboration of studies, management of natural resources, planning of the electricity system.
Autonomous Region of Príncipe (ARP) **• Regional Secretariat for the Environment and Sustainable Development.**	Regional Administration of Príncipe Island, responsible for the implementation of national regulations, in health, environment, agriculture and fisheries.
Districts/Local Authorities	Responsible for public lighting, supplied by the low voltage distribution network.
General Regulatory Authority (AGER)	Regulation of the electricity sector.
Water and Electricity Company (EMAE)	EMAE is the utility company responsible for generating, transmitting, and distributing electricity.
National Petroleum Agency (ANP)	Public regulatory and promotion entity for the oil and gas sector.
Fiduciary Agency for Project Administration (AFAP)	Responsible for managing São Tomé and Príncipe Government projects financed by the World Bank.
Coordination Committee for the Electricity Sector Transformation Program (CC-PTSE)	Responsible for supporting the government in the reform of the electricity sector.

are still very few companies that include renewable energy in their service portfolio, due to the small size of the market and the lack of openness to private sector participation. Only the following entities were identified: Horizonte Electricidade, Tecnologia e Serviços Lda, NGO TESE Association for Development, and RENERGIE Lda."

However, the report adds, "In recent years, there has been a very interesting phenomenon in São Tomé and Príncipe, the creation of not only one but two national renewable energy associations, namely the Santomense Association of Renewable Energies (AENER) and the Association for the Promotion of Renewable Energies and Sustainable Environment of São Tomé and Príncipe[16] (APERAS)."[17] Both associations aim to raise awareness within the community of São Tomé and Príncipe to sustainability issues, particularly the use of renewable energies in the country.

The above description of projects previously carried out or in progress reveals a strategy for addressing the problems related to the energy sector from international cooperation with experienced institutions dedicated to development support. These include The African Development Bank (AfDB), The Arab Bank for the Economic Development of Africa (BADEA), The

European Investment Bank (EIB), The World Bank Group (WB), the European Union (EU), The International Monetary Fund (IMF), The United Nations Industrial Development Organization (UNIDO), and The United Nations Development Programme (UNDP). Furthermore, there is still a demand for establishing bilateral relations with Angola, Brazil, Equatorial Guinea, Libya, Morocco, Nigeria, Portugal, and the People's Republic of China. These partnerships provide the technical and financial support necessary for the development of São Tomé and Príncipe but have been insufficient.[18]

The government of São Tomé and Príncipe has assumed this cooperation strategy. According to the Minister of Infrastructure, Natural Resources, and the Environment, Osvaldo Abreu,[19] the country's current (2022) electricity consumption varies between 18 and 22 megawatts. However, by 2025 consumption will almost triple to 50 megawatts, and by 2030, it will increase to about 100 megawatts. The Minister stated that the country spends more than 25 million euros yearly on fossil fuel for electric generators in three thermal power plants, plus the expensive maintenance (2 to 3 million USD a year), which is unaffordable for a small country with a fragile economy. São Tomé and Príncipe's government plan is to ask for support from the World Bank and construct at least three hydroelectric plants and a photovoltaic production network, in a global investment of around 100 million USD. In the interview, Osvaldo Abreu clarified that the necessary investments were to be made based on international cooperation by saying:

We have well-established partnerships with international organizations, namely the World Bank, different United Nations agencies, oil companies, and other international institutions linked to renewable energy, including EDP renewable from Portugal, which in addition to being interested in developing solar energy production projects on the island of Príncipe, has also contributed to the development of exploratory studies on the country's capacity to produce this type of energy.

According to the Minister, these talks resulted in commitments with companies from several countries, namely Portugal, Germany, Spain, the United States of America, and Brazil.

To mitigate the intermittent production of photovoltaic electricity, the government intends to construct several hydroelectric plants costing over 50 million USD. The government intends all these renewable energy projects to be funded by private investment using the modality Build, Operate, and Transfer (BOT) which eliminates the need for initial investment by the contracting party. Three new dams for electricity production are planned for São Tomé (where most of the country's population resides), one on the river

Yô Grande, located in the south of the island, another on the river Lembá in the north of the island and a third on the river Abade, also in the south. A fourth is planned on the river Xufi-Xufi on the island of Príncipe, where several micro-production dams will be built. The Minister also mentioned the need for further investments, such as storage batteries, to safeguard stable electricity distribution.

6. CONCLUSION

This chapter examined the outlook for energy transitions in São Tomé and Príncipe, a small island developing state in Africa, across various dimensions, including the country's energy profile, national policies, institutions, and emerging challenges and opportunities.

São Tomé and Príncipe (practically located on the equator) enjoys rainy seasons, with plenty of water to fill dams for hydropower production. The envisioned steady hydroelectricity production, assisted by photovoltaic plants, which could eventually be reinforced by wind and wave or even geothermal energy sources, could solve most of the net zero energy requirements of São Tomé and Príncipe by 2050.

As in other SIDS, the significant restriction might be financial. However, São Tomé and Príncipe seems to have found a way to solve that by presenting its project as a business opportunity, attracting private investors.

Due to the disappointment of (still) not being an oil-producing country, rising fossil fuel costs, the growing restrictions on receiving refined oil products from Sonangol (Angolan partner company), and the concerns with global warming, the country started an ambitious dam and photovoltaic park building project, to distribute electricity to its population, in a better, cheaper, and autonomous mode. The project is mainly driven by opening São Tomé and Príncipe's electricity market to private investment with a Build, Operate, and Transfer model, attracting interested international partners.

These circumstances and the resolved central weaknesses of São Tomé and Príncipe (the financing of its intervention and the organizational structure) will lead to positive externalities due to the promotion of renewable energy, energy efficiency, and actions favorable to the planet's sustainability. From an economic point of view, São Tomé and Príncipe will reduce its dependence on fossil fuel imports, thus freeing up resources for other sectors and, simultaneously, generating opportunities for productive energy use that will create employment and an economic boost. Universal accessibility to clean energy, particularly in households, will be a step toward a better quality of life, with improved health and education. Furthermore, the increased accessibility of clean energy will reduce the demand for forest materials and the emission of greenhouse gases.

In summary, São Tomé and Príncipe possesses several strengths that make it well-placed for the upcoming energy transitions. These include a strong political commitment to combating global warming, key global and regional organizations membership, and its location near the equator, which provides optimal solar irradiance. Additionally, the country is secure, receives significant international support for the energy sector, and has promising prospects for wave and hydropower generation. Lastly, São Tomé and Príncipe is restructuring its national electricity entity to attract foreign investment in renewable energy.

Despite its strengths, São Tomé and Príncipe faces several challenges to its energy transition. As an impoverished country, approximately 30% of the population lacks access to electricity. The population is also unevenly distributed, with 75,8% concentrated in urban areas and 25% living in the Capital, São Tomé city. The other 25% of the population is dispersed. Moreover, the country heavily depends on external financial support and investment, and its energy distribution grid suffers significant losses. It is also perceived as an unattractive investment destination, leading to a slow transitional process largely dependent on third-party actors. The country also has organizational weaknesses, including management skills and lack of qualified human resources.

Nevertheless, São Tomé and Príncipe has multiple opportunities to advance its energy transitions. The recent increase in fossil fuel costs will accelerate the deployment of alternative energy sources. The country has the potential to attract international investments to construct four hydropower generation dams. Reduced energy costs could catalyze other industries, such as tourism, maritime economy, agriculture, and water scarcity solutions through desalinization, promoting social development. Restructuring the water and electricity company (EMAE) could facilitate private and foreign investment. Existing technology for creating renewable energy mini-grids may address the current grid's efficiency problems.

Considering the threats and dangers that could hinder São Tomé and Príncipe's energy transitions is crucial. Global warming could cause dramatic weather effects in the Gulf of Guinea, impacting the country's renewable energy potential. Additionally, while São Tomé and Príncipe is a stable and secure country, the Gulf of Guinea region faces security challenges that could spill over and affect the nation's ability to manage its energy transition. The country's fragile position makes it crucial for the international community to support its efforts and mitigate potential threats.

Several lessons can be extracted for energy transitions in small island developing states:

1. Political commitment and policy support: A strong political commitment and appropriate policies are essential to the success of energy transitions;
2. International cooperation: Given their often-fragile economies and multiple internal vulnerabilities to external shocks, it's crucial for SIDS to seek and nurture international support. This can help mitigate potential threats and accelerate the transition process;
3. Attract private investment: São Tomé and Príncipe used a Build, Operate, and Transfer model to attract international partners, demonstrating that innovative financial models can effectively overcome financial constraints by drawing in private investment;
4. Prepare for challenges: SIDS should anticipate and prepare for potential threats and challenges. These includes the need for capacity building in areas such as project management and technical skills training.

In a nutshell, the experiences of São Tomé and Príncipe highlight the importance of strategic planning, leveraging unique natural resources, engaging private investors, and securing international support for a successful energy transition in SIDS.

REFERENCES

African Union Energy Commission. (n.d.). *Sao Tome and Principe Energy Balance 2020*. AU AFREC. https://au-afrec.org/en/central-africa/sao-tome-and-principe

ALER. (2020). *Renewable Energy and Energy Efficiency in São Tomé and Príncipe - National Status Report.* ALER. https://www.lerenovaveis.org/contents/lerpublication/aler-relatorio-stp-nov2020.pdf

CIA. (n.d.). *Sao Tome and Principe - The World Factbook.* CIA. https://www.cia.gov/the-world-factbook/countries/sao-tome-and-principe/

IRENA. (2019). *A new world: the geopolitics of the energy transformation*. IRENA.

IRENA. (2023). *SIDS Lighthouses Initiative Annual Progress Report*. IRENA. https://mc-cd8320d4-36a1-40ac-83cc-3389-cdn-endpoint.azureedge.net/-/media/Files/IRENA/Agency/Publication/2023/May/IRENA_SIDS_LHI_progress_2023.pdf?rev=6aac8f77eede4b768a078cc4a971c543

Leandro, F. J. B. S., Martínez-Galán, E., & Gonçalves, P. (2023). *Portuguese-speaking Small Island Developing States*. Springer Nature Singapore. doi:10.1007/978-981-99-3382-2

Ministério das Infraestruturas e Recursos Naturais. (2021). *Plano de Ação Nacional no Sector das Energias Renováveis*. Ministério das Infraestruturas e Recursos Naturais.

República Democrática de São Tomé e Príncipe. (2019). *Plano Nacional de Desenvolvimento Sustentável de São Tomé e Príncipe*. Republica Democratica. https://financas.gov.st/phocadownload/Planeamento/publicacao/Plano%20Nacional%20de%20Desenvolvimento%20Sustentavel%20-%20STP%20-%202020-2024.pdf

Simões, J., Leandro, F. J., de Sousa, E. C., & Oberoi, R. (Eds.). (2023). *Changing the Paradigm of Energy Geopolitics*. Peter Lang Verlag., doi:10.3726/b18776

United Nations Economic Commission for Africa. (n.d.). *African small island developing states*. UNECA. https://archive.uneca.org/africansmallislanddevelopingstates/pages/african-small-island-developing-states

UNOSSC. (2021). *Good Practices in South-South and Triangular Cooperation for Sustainable Development in SIDS*. UNOSSC. https://unctad.org/news/small-island-developing-states-face-uphill-battle-covid-19-recovery

World Bank. (2023). *Leveraging Natural Wealth to Build Opportunities*. World Bank. https://documents1.worldbank.org/curated/en/099457104212338173/pdf/IDU0d4e660690db44049cc0bd440699b9c1b4544.pdf

World Bank. (n.d.). *Sao Tome and Principe Overview: Development news, research, data*. World Bank. https://www.worldbank.org/en/country/saotome/overview

ENDNOTES

1. The United Nations Office of the High Representative for the Least Developed Countries, Landlocked Developing Countries and Small Island Developing States (UN-OHRLLS) supports 90 vulnerable Members States including 46 Least Developed Countries (LDC), 32 Landlocked Developing Countries (LLDC) and 37 Small Island Developing States (SIDS). Some countries in these last two groups also belong to the LDC.
2. Retrieved from https://www.un.org/ohrlls/sites/www.un.org.ohrlls/files/good-practices-in-sstc-for-sustainable-development-in-sids-web.pdf
3. Retrieved July 20, 2023, from https://www.anp-stp.gov.st/index.php/pt/publicacoes/noticias-pt/item/175-conclusao-da-perfuracao-do-poco-jaca-1
4. Retrieved July 20, 2023, from https://thesouthernhub.org/
5. Retrieved July 20, 2023, from https://data.worldbank.org/indicator/EG.FEC.RNEW.ZS?locations=ST
6. Green Climate Fund (GCF) – a critical element of the historic Paris Agreement - is the world's largest climate fund, mandated to support developing countries raise and realize their Nationally Determined Contributions (NDC) ambitions towards low-emissions, climate-resilient pathways
7. Retrieved April 20, 2023, from https://www.afdb.org/pt/news-and-events/press-releases/sao-tome-et-principe-la-banque-africaine-de-developpement-soutient-la-reforme-economique-et-le-secteur-de-lelectricite-avec-un-don-de-plus-de-6-millions-deuros-35650
8. Retrieved April 20, 2023, from https://www.aler-renovaveis.org/contents/activitieseventsspeakersdocuments/ceutonia-lima-neto_1117.pdf
9. According to PANER, it is: "Supply of parts for generators; Construction of a photovoltaic sub-station at the Sto Amaro Power Station (1.5 MW) (complementary to UNDP); Rehabilitation of the Papagaio water plant; Construction of a cargo terminal and supply of specific equipment".
10. Retrieved April 20, 2023, from https://projects.worldbank.org/en/projects-operations/project-detail/P157096
11. The ECCAS region is facing similar economic, social and environmental challenges as other regions in Sub Sahara Africa. ECCAS is comprising eleven member states whereas most of them are classified as low or lower-income countries. ECCAS countries represent a growing population of 172 million and remain fragile in a political and economic view. In the energy sector, the region faces a bundle of challenges. Although the

region has vast fossil fuel and renewable energy potentials, the access rate to modern, reliable and affordable energy services remains very low. The total electricity production and consumption is lower than in other African regions.

12 Retrieved July 20, 2023, from https://www.gn-sec.net/content/centre-renewable-energy-and-energy-efficiency-central-africa-cereeac

13 *idem*

14 Retrieved July 20, 2023, from https://www.gn-sec.net/content/background

15 *idem*

16 AENER has planned the realization of some projects, such as: Feasibility study of wind energy in the South Zone of the Island of São Tomé; Feasibility study of the installation of solar panels in the North Zone of São Tomé Island; Project to raise awareness of children in primary schools about the use of energy; Pilot project to feed a small community with alternative energies; Project to raise awareness of the population to control energy consumption; Project to hold conferences and debates on the subject of renewable energies.

17 APERAS is a private, non-profit and national association whose mission is to promote, sensitize, educate, train and empower São Tomé society on the best practices for the development of the national energy sector, its integration with the protection of the environment and biodiversity, as well as the mitigation of the effects of climate change.

18 The several UN agencies present in country have provided São Tomé e Príncipe 22,823,609 USD of which 6,058,531 were dedicated to Social Cohesion, 6,811,869 USD were for Good Governance and 9,938,210 USD for Economic Growth and Resilience where the climate change aspects are mainly dealt with. The UN system itself was sponsored by many entities and those entities themselves have also engaged in bilateral cooperation with the Government, in order to certify that their contributions were well taken care off. Australia; India; Universities (University of California and University of Edinburgh); Portugal; Japan; Azerbaijan; Denmark; Austria; Canada; Global Partnership for Education; Korea; European Union; Global Environment Facility (GEF); Private sector companies; World Bank; SDG Fund of the UN SG Secretariat; The Global Fund for AIDS, Tuberculosis and Malaria; African Development Bank; Germany; Rural Poor Stimulus Facilities; UK and USA. The United Nations 2021 Annual Results Report on São Tomé e Príncipe clearly states that the 2030 Agenda for sustainable development (UN Sustainable Development Goals) for São Tomé e

Príncipe "requires a substantial increase in investment and the various sources of financing for the SDGs"; public finance and foreign resource inflows (ODA was 12.3% of GNP in 2020 while FDI amounted to 9.96% of GDP, compared to 1.9% of GDP for bank loans) dominate Sao Tomé and Principe's financial landscape; remittances from the diaspora stayed low (1.82% of GDP in 2020); only 48.9% of the state budget is funded domestically, while 51.1% of it is funded through grants and loans; approximately 95% of the budget for public investment is financed by outside sources, compared to 5% by locally produced sources.

19. Retrieved May 20, 2023, from https://mercadosafricanos.com/sao-tome-e-principe-investe-forte-nas-renovaveis

Chapter 5
A Framework of Sustainable Economics:
Determining Driving Forces of the Sustainable Development in the MENA Region

Nima Norouzi
https://orcid.org/0000-0002-2546-4288
Bournemouth University, UK

ABSTRACT

The Middle East and North Africa region has become one of the most important regions globally due to its energy reserves and geopolitical position. International organizations and researchers have discussed the challenge of sustainable development in the MENA region for many years. In this chapter, by examining the causal relationships between the four components of sustainable development, it is tried to understand better the interaction between sustainable development components in the region. Therefore, by collecting 92 institutional, bio-environmental, economic, and social variables, development indicators were constructed using the principal component analysis method for 2000-2020. This relationship was then tested using a Granger causality model and a dynamic data panel model. The results of this study show well that the economic development achieved at the expense of environmental degradation in this region failed to improve the non-economic components of sustainable development and provided the basis for their decline.

DOI: 10.4018/978-1-7998-8638-9.ch005

INTRODUCTION

The Middle East and North Africa region, known as "MENA," has become one of the most important regions globally due to its energy reserves and geopolitical position as a link between Asia, Europe, and Africa (Anderson & Hsiao, 1981). From the point of view of economic structure, the countries of MENA can be divided into three groups: A. Countries that do not have natural resources but have a large labor force (such as Djibouti, Egypt, Jordan, Lebanon, and Tunisia). B. Countries rich in natural resources have abundant labor (such as Algeria, Iran, Iraq, Syria, and Yemen). C. Countries that are rich in resources but lack labor (such as Bahrain, Kuwait, Libya, Amman, Qatar, Saudi Arabia, and the UAE), so it can be said that the economic development of some of these countries is dependent and dependent on materials. Others depend on their agricultural sector(Apergis, 2004).

Despite the abundance of natural resources, the countries of this region have always suffered from a common pit, which has created serious obstacles to their sustainable development. Military and political instability is the first common feature and obstacle to development in the MENA region(Arellano & Bond, 1991). The strong military presence in countries such as Egypt, Algeria, and Tunisia, the rule of ethnic minorities in the Gulf countries, and the oil war in the region are examples of this instability(Atukeren, 2007). The second factor in underdevelopment is the large dependence of some countries on oil revenues, which has led to rent-seeking behaviors, strengthening bureaucracy, and limiting and weakening democratic institutions (Baltagi, 2021). Another feature of the region is the special colonial heritage in the region and the greed of world powers to dominate the vast hydrocarbon reserves of the region, which has led to the skepticism of some dependent governments and thus the democratization and restriction of democracy.

Although the reform process in the region is slow, the improvement of the education and health system, the expansion of women's participation in some countries of the region, the reform of some countries to move towards good governance, and the availability of new generations will improve. Institutional, environmental, economic, and social components are sustainable development in this region. In this paper, by examining the causal relationships between the four components of sustainable development, we try to understand better the interaction between sustainable development components in the region over the past years(Bhattarai & Hammig, 2001). Therefore, by collecting 92 institutional, bio-environmental, economic, and social variables, development indicators were constructed using the principal component analysis (PCA) method. The results of this study clearly show that the economic development achieved at the expense of environmental degradation in this region has failed to improve the non-economic and non-economic components of non-economic development(Booth, 1995).

The present chapter consists of four sections. After the introduction and in the second part, the theoretical foundations of research and review of past studies are discussed. The third part expresses the structure of the Granger causality model, defines the variables, extracts the horns using the principal component analysis method and expresses their process, and finally tests the significance. The fourth section presents the results of the Granger causality test, and finally, the fifth section is devoted to summarizing and concluding.

BACKGROUND

Simultaneously with the Chen & Feng (1996) publication and the subsequent agenda of the 21st Rio Summit, a wave of global concerns about the state of sustainable development swept the world's countries, especially the developed countries. In the meantime, many researchers focused on the two dimensions of economic and bio-environmental development and the factors that shape or influence them. Economic growth was considered a cause and environmental pollution and analysis of natural resources as a disadvantage (Costantini & Martini, 2010). These studies can be included in the framework of the Kuznets Environmental Curve (EKC) hypothesis. Inspired by the inverted U-curve of the Kuznets curve, these studies believe that with increasing economic growth, environmental pollution increases, initially, and decreases after passing through a stage. Dinda (2004) and Stern (2004) provide a comprehensive overview of the studies carried out in the hypothetical field of the Kuznets bioenvironmental curve in a comprehensive review. Numerous empirical studies, such as those of Diao et al. (2009), can be found to confirm this hypothesis. The types of pollutants and bioenvironmental hazards are not true, and the results are sensitive to the temporal and geographical area under study(Chen et al., 2019; Desai et al., 2005).

Over time, researchers turned their attention to other dimensions of sustainable development, including sustainable institutional development. Researchers such as Piñeiro Chousa et al. (2017), Culas (2007), Bhattarai & Hammig (2001), and Tamazian et al. (2009) have tried to add institutional development variables as explanatory variables in their experimental model. Firstly, the relationship between economic growth (per capita income growth) and environmental analysis remains stable, and secondly, these variables, by taking negative coefficients, are practically able to mitigate the growth rate with their development. They believe that in addition to the two-stage relationship between economic growth and environmental development as a cause and effect, institutional development can also, with its growth, bring about the consequences of environmental development. Studies have also been conducted in the field of decision-making (democratic or non-democratic) and its effect on environmental development, including Farzin and Bond (2006), Menegat (2002), and Norouzi & Dehghani (2020). However, there is a positive cause and effect between democracy (as institutional development) and environmental development; Norouzi & Kalantari (2020) argues that the relationship between democracy and the environment is not necessarily conceivable.

Extensive studies can also be found that have examined the effects of institutional development on other components of sustainable development, including economic development. Olson (1993) is one of the pioneers in this field who believes that the dictator, due to his short life, Unable to comply with economic principles such as respect for property and contracts, strives to maximize his interests and those of those around him in this short period by underestimating the principles of development. And economic growth. Chen & Feng (1996) show that instability, polarization, and political repression negatively affect economic growth. Norouzi et al. (2020) and the papers of Sachikonye (2002), Heller et al. (2007), Joseph (2001), and Söderbaum (2007) also emphasize the importance of developing and deepening democracy towards economic development.

In contrast, OECD (2008), by presenting a data panel model consisting of 122 countries, shows the growth rate of income, the level of primary income, and the political system political stability or instability significant effect. Desai et al. (2005) also show that the relationship between social and economic variables (inequality and inflation) depends on the political structure of society. In his interesting study, Newman and Thomson (1989) categorized the studies on the causal relationship between institutional (social) development and economic development into four groups: A. Social development as The product of economic growth, B. Economic growth as a product of social development, C. Endogenous and close relationship between economic growth and social development (Frederking, 2002) and the lack of existence. Finally, their research study in the form of a data panel model confirms a causal relationship between social development and economic growth.

METHODS AND MATERIALS

Method Mathematics

Equation (1) shows the general structure of the Granger causality model in the form of data panel models. In this regard, Y is the disabled variable, X is the causal variable, Z is the control variable, ϕ, γ, β are the coefficients of the explanatory variables, v is the waste component, and η is the intangible variable, which represents the specific characteristics of each country (Costantini & Martini, 2010; Narayan et al. 2008; Lee et al., 2008; Norouzi, 2019)

$$Y_{it} = \sum_{l=1}^{m} \beta_l Y_{it-1} + \sum_{l=1}^{m} \gamma_l X_{it-1} + \sum_{l=1}^{m} \varphi_l Z_{it-1} + \eta_i + v_{it} \quad (1)$$

As mentioned, the variables studied in this study are indicators of sustainable development for institutional, environmental, economic, and social components. Due to the lack of comprehensive indicators that clearly show each component's dimensions, these indicators are useful. The chapter authors developed the principal component analysis method, briefly discussed in the following sections(Norouzi & Fani, 2021a).

Given the existence of these four indicators, there are practically 21 possibilities for a causal relationship between indicators, but Norouzi (2020), in his article, showed that it is necessary to get rid of false causality errors in the Granger causality test, plus Indirectly distinguish from direct causality. This is the present study means that for each of the above 21 cases, three equations must be designed. For example, if we are looking to test the causality of an institutional index over an economic index, we must The control of the environmental index is tested again with the social index and the third time with both indices as control variables. Therefore, by shifting the indices, in practice, 63 equations such as equation 1 were tested. Finally, the causal relationship of the variable X to Y can only be confirmed if, in all three cases, the coefficients γ are significant and the hypothesis of zero is based on the sum of the coefficients γ (parent test) is rejected(Norouzi & Fani, 2021b).

Variables and Framework

Most of the studies mentioned in the analysis of causality among the components of sustainable development have used alternative variables instead of development indicators. Environment, per capita income growth as indicators of economic development, child mortality, life expectancy as indicators of social development, and political stability as indicators of institutional development have been widely used. The disadvantage of these variables is the lack of comprehensiveness and sufficient comprehensiveness of the mentioned development component. Therefore, in the present study, an attempt has been made to create more comprehensive indicators for each development component using the principal component analysis method(Justesen, 2008; Norouzi & Fani, 2020).

The principal component analysis is a statistical method that has been developed to reduce the dimensions of a set of related data. The output data of this method, known as principal components (1PCs), are linear combinations of primary data and non-correlated vectors that are ranked based on the explanatory variables of the original data. Although this method produces the principal component by the number of variables in the primary data set, the first few principal components can explain more than 90% of the variance of the primary data (Jolliffe, 2002; OECD, 2008).

To begin the analysis, it was first necessary to summarize the constituent variables of each domain, for all countries in the MENA region, over some time under analysis. Among the countries in the MENA region, three countries, Afghanistan, Iraq, and Syria do not have a statistical system due to numerous wars and internal crises, so comprehensive databases such as the World Bank database did not have the information we needed these countries. Therefore, these countries were excluded from the spatial scope of the study. To correctly understand how the trend of sustainable development indicators is changing, the period from 2000 to 2020 was considered. Finally, a set of 92 variables was collected, which is summarized in Table 1.

Criteria and Indicators

A look at Table 1 shows that a great deal of time and attention has been spent in gathering information and a variety of variables. As it turns out, the above data are composed of many different scales. Therefore, it is necessary to perform two basic processes on the raw data to start the analysis. At first glance, some data are exponential, and some are declining, meaning that an increase in some indicates a sustainable development improvement, and some indicate a decline in development in the country in question.

Such as GDP per capita or the total cost of education and health are among the exponential variables that show a higher level of sustainable economic or social development, but on the contrary, indicators such as political rights and rates Ginny are degenerative variables; That is, an increase in them indicates a decrease in development in the country. Therefore, it is necessary to first have all the variables in one direction. For this purpose, all degradation data is multiplied by a negative until they become incremental data. The second change required is the elimination of units, which is done during the normalization process. For this research, the scale normalization method is used again, which is shown in Equation 2(Shahbaz et al., 2018).

$$Re-scaled-value=\frac{x-x_{min}}{x_{max}-x_{min}} \tag{2}$$

In this way, the scale of all the data is lost, and all the data of each variable are re-scaled from zero, with the value of zero belonging to the least data and the value belonging to the most data. The principal component analysis was performed for 61 countries in the MENA region by applying the previous two steps. After obtaining the principal components (PCs) for each of the institutional, environmental, economic, and social developments, they were originally put together to explain more than 09% of the variance of the original data. Therefore, to calculate the indicators of sustainable institutional development of the five main components, environmentally sustainable development, four main components, sustainable economic development of the five main components, and sustainable social development, five main components were added to each. The four dimensions of sustainable development come to the fore.

The Trend of Sustainable Development Indicators in the MENA Region, Problems

By calculating the four indicators of sustainable development, we can look at the trends of these components in the time frame of the research (2000-2020). Figure 1, which shows the above indicators for all countries in the region, clearly shows that the only component grown in this period is economic development.

Table 1. Variables that constitute the components of sustainable institutional, environmental, economic, and social development

Criteria	Scale	Sub-criteria or indicator	Reference
Constitutional Sustainability	Qualitative, [1-7]	Political rights Freedom of citizenship	Freedom House Organization (freedomhouse.org) Reporters Without Borders(www.rsf.org) Kaufmann et al. (2009), General and Individual Indicators of Shahbaz et al.(2020), World Bank
	Qualitative, [0,inf]	Press Freedom Index	
	Qualitative, [-2.5,2.5]	Criticism and accountability Political stability and the absence of violence and terrorism Government efficiency Quality of legislation The rule of law Control of corruption	
Environmental Sustainability	Quantitative, %GNP	Damage caused by carbon dioxide Reduction of energy resources Reduction of minerals Analysis of forest reserves Damage caused by particulate matter	Global Development Indicators, World Bank, Different Years
Economic Sustainability	Quantitative, $	GDP per capita (PPP), proven compared to 2000	Global Development Indicators, World Bank, Different Years
	Quantitative, %	Per capita growth of GDP	
	Quantitative, %	Inflation rate (consumer price)	
	Quantitative, %GDP	The cost of starting a business	
	Quantitative, %GNP	Current account balance	
	Quantitative, %GNP	Foreign direct investment, net inflows	
	Quantitative, of 100 people	Internet users	
Social Sustainability	Quantitative, %	Gini coefficient	Global Development Indicators, World Bank, Different Years
	Quantitative, %GDP	Total health costs	
	Quantitative, Subscribers from the whole population	Access to treated water resources	
	Quantitative, year	Life expectancy at birth	
	Quantitative, of 1000 people	The mortality rate of children under five years	
	Quantitative, %	The share of women in the seats of the National Assembly	
	Quantitative, % of GDP	The total cost of education by the public sector	
	Quantitative, %	The ratio of girls to boys in primary and secondary education	

Figure 1. Sustainable social, economic, environmental, and institutional development trends in MENA region (2000-2020)

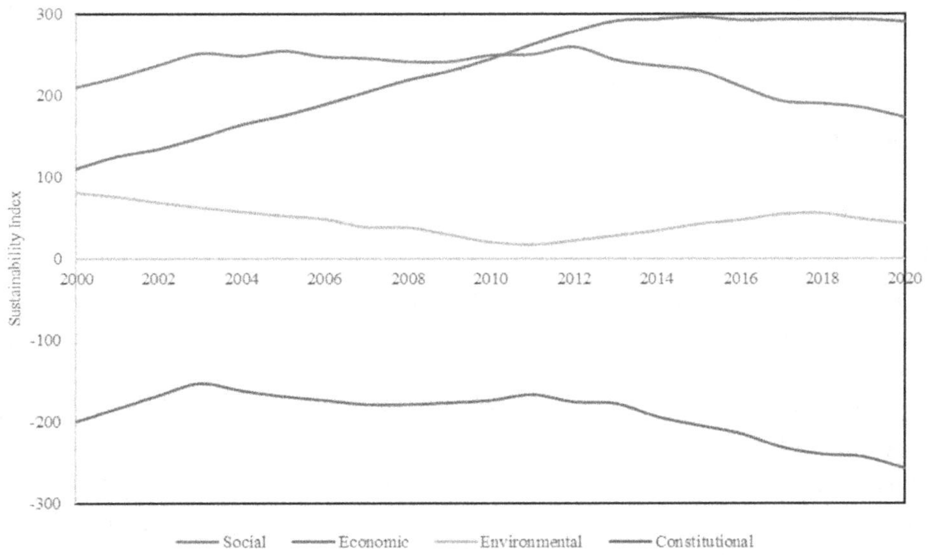

The MENA Environmental Development Index, which had a positive figure in 2000, declined over time and eventually turned negative at the end of the eight years. The other two institutional and social development indicators also decrease to a level less than the initial level after the ups and downs. The outbreak of war in 2001 between international forces and the Taliban in Afghanistan (which eventually led to the occupation of Afghanistan) and the war in Iraq in 2003, and the fall of the Ba'athist regime are two major milestones in the region during the investigation period. The destruction of infrastructure in the target countries, the killing and displacement of millions of people, the severe destruction of the environment, and the disintegration of government institutions during the fall are the obvious consequences of any war. In addition to the above wars, the spread of terrorism, the increase in the production and transit of narcotics, many such factors can be mentioned to destroy institutional, social, and environmental components in the MENA region. Of course, other cases include the Syrian civil war and the coup in Egypt, which were among the most vital issues of the year (2012-2013), and the emergence of the Islamic State of Iraq and the Levant or ISIS (2014). The improvement in the economic component of the research period is due more than anything else to the increase in the price of oil and oil products in the world. However, in addition to rising oil prices, some countries in the region, such as Turkey, the United Arab Emirates, and Saudi Arabia, need to properly invest in infrastructure, industry, commerce, and tourism to improve their economic performance. And, the oil market suffered severely in the wake of the Covid-19 crisis and imposed heavy costs on the MENA Basin countries, in addition to their public health costs(2019-2021).

Statistical Tests

Before beginning the Granger causality test, it is necessary to ascertain the significance of the extracted horn. One of the most important weaknesses of experimental studies, especially in the field of data panel models, is the lack of commitment of researchers to test the materials used in the model for example, in a study such as the Kónya (2006) study on the causal relationship between exports and growth, Hartwig (2010) on the causal test between economic freedom and economic growth, Hsiao & Hsiao (2006) on income-related relationships. And environmental pollution and Hoffmann et al. (2005) in the causality test between foreign direct investment and pollution that all used data panel models to examine the causal relationship, a sign of mania test or even a reference to it doesn't exist. This is especially true for non-economic researchers(Stern et al., 1996; Holtz-Eakin et al., 1998).

Table 2. Test results for four indicators of institutional development, environmental, economic, and social

Test	Constitutional	Environmental	Economic	Social
Levin, Lin & Chu t	-7.33(0.00)	-31.51(0.00)	-31.52(0.00)	-10.22(0.00)
Breitung t-stat	-	-0.099(0.53)	-7.33(0.00)	-0.98(0.15)
Im, Pesaran and Shin W-stat	-1.88(0.03)	-12.31(0.00)	-6.59(0.00)	-1.09(0.14)
ADF - Fisher Chi square	46.31(0.01)	170.71(0.00)	150.49(0.00)	49.84(0.00)
PP - Fisher Chi-square	39.91(0.06)	223.16(0.00)	248.91(0.00)	105.11(0.00)

Note: Test has been performed in three environmental, economic, and social indicators with the assumption of width from the source and trend variables and the institutional index only with the assumption of width from the origin. The numbers in parentheses indicate the probability of accepting the null hypothesis(Miljkovic & Rimal, 2008).

Hsiao (1982) has provided a complete description of mana test methods in the data panel model. Tests such as Levin and Lane, Ime, boys and Shane, Hadry test, and Fisher tests are among the mana test tests in data panel models with different assumptions and even zero hypotheses. Table 2 summarizes the results of the Mana test for the four indicators of institutional development, environmental, economic, and social. Hypothesis zero in the tests of this table is the existence of a single root. This table shows the rejection of the null hypothesis and thus the significance of the indicators.

Selecting the Optimal Intervals

Since the causal analysis and its results are strongly dependent on the number of explanatory endowments, it is necessary to know the optimal extent of these endowments before implementing the model. In their articles, Miyakoshi & Tsukuda (2004) and Justesen (2008) mention three methods in selecting the optimal intervals. In the first method, the intervals are entered into the model without any statistical test and only based on the researcher's opinion. This method is based on the individual supervision of the researcher, and in practice, a model in different researches may be specified in different ways and easily achieve different results. In the second method, statistical indices such as the Akaike Information Index (AIC) or the Schwartz Index (SC) indicate that the maximum endpoints are optimal(Miyakoshi & Tsukuda, 2004). This method is also problematic, especially in models with short time intervals, because with each interruption, one degree of freedom of the model decreases. In the third method used in this article, a combination of the above two methods is used. In this way, the researcher determines the maximum interruptions based on his knowledge and the time interval of his model, and then using the Akaik index and, Determines the best amount of interruption in this pre-determined play. In the present study, due to the small chronological recurrence (8 years) and the model implementation method, a maximum of two interruptions was considered for the explanatory variables. Two breaks are the best amount of breaks, and in each remaining relationship, one break will give the best results.

RESULTS AND DISCUSSION

The last step in any experimental research is to implement the model and extract the results. The implementation of the Granger causality data model has been accepted in four different ways and with different assumptions. In the first method, the model of estimation assuming fixed effects (FEE) is practically used, in which the variable η is eliminated by mediation. Applying both of the above methods to the implementation of Equation 1 yields biased results because, in both methods, the relationship between the intermittent components of the interdependent variable and the remainder of the dependent component is negligible. For example, Anderson & Hsiao (1981) suggests eliminating this problem, and the invisible variable should be removed first by multiplication and then by using $\Delta y_{i,t-2}$ or $y_{i,t-2}$ dependent on c, t with y. The correlation between it and the partial differential of the waste is practically lost, and this method, although it is possible to estimate the compatibility of the model coefficients, is not necessarily efficient (Baltagi, 2021). The latter method eliminates the previous problems of accumulation of the Panel-Data's data, proposed by Holtz-Eakin et al. (1988) and Arellano and Bond (1991). Using instrumental variables, which are, in fact, explanatory variables with two or more intervals, the model is implemented by the GMM method. Residues are extractable results (Hartwig, 2010). Figure 2 shows the results of Equation 1 for the 63 equations mentioned above. The results are shown using the one-step method and the two-stage Arellano and Bond method.

Figure 2. The causal relationship between institutional, environmental, economic, and social components of sustainable development in the Middle East and North Africa (MENA) region

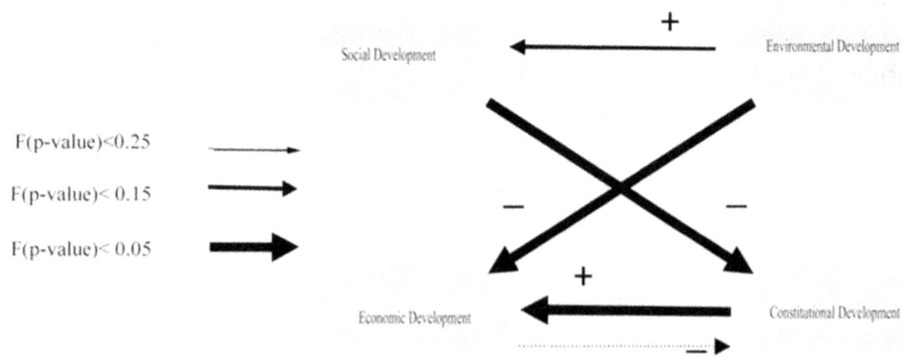

Regarding sustainable environmental development, it should be said that despite the causal impact of this component on the two components of sustainable economic development and sustainable social development, changes in this component are not the result of any of the other components. Natural resources, whether in the form of hydrocarbon reserves or the form of land, climate, etc., are the most important sources of economic growth for the countries of the MENA region. Extensive use of these resources and unlimited extraction of underground resources in the region has led to a negative causal relationship from environmental development to economic development. This means that economic development in the MENA region comes at the expense of environmental analysis and degradation. Another effect of this component is on social development, which seems to affect health the most. Therefore, it can be seen that the extraction and analysis of natural resources in the countries of the MENA region over the past decades have led to increased economic development and reduced social development.

Institutional development, which means increasing and deepening democratic work and improving good governance, is highly capable of increasing economic development. Interestingly, economic development over the research period has an adverse, albeit weak, effect on institutional development. This result may seem vague at first glance, but considering countries' experiences in the region can well determine this relationship. A good history shows that the region's rulers have increased their oil revenues, above all, by strengthening their power bases and b Expression in their countries. This shows well the inverse impact of economic development on institutional development.

Examining the economic component of sustainable development in the MENA region can answer many questions about the effects of rising oil revenues. As it is clear, the prevailing pattern in the region shows that the rise in oil prices and the sudden influx of oil revenues to the countries of the region, in general, has not been a driving force for any of the other components of sustainable development. The recent experience of our country is also a confirmation of this experience. The only causal effect of economic development is limited to the same negative effect of this component on institutional development.

As shown in Figure 2, the only channel of influence for social development is institutional development. But this relationship is also as negative as the causal relationship between economic development and institutional development. When seeing this relationship, the question that arises for any reader is whether social development, which means increasing education, health, or women's participation, should improve institutional development. Although the experimental results of the present model show a negative causal effect of social development on institutional development, this finding is certainly the beginning of further research in this regard. It should be noted that the above negative causal relationship has been achieved in a period between the period (eight years) and only among 61 countries in the MENA region. Whether is time-sensitive and geographically sensitive or whether the above relationship is conditional on the political or economic structure of countries are questions Perhaps in the present position, it can be hypothesized that the positive impact of social development on institutional development requires appropriate channels of influence such as democratic institutions or governments of the will of the people, although these paths of influence in most parts of the country. MENA is not available, or it works very inefficiently.

Following the goal of the research, the following results have been obtained:

- The main tasks for a new scientific direction for the transition period of creating the green economy have been formed;
- The definition of the concept of the "green" economy is formulated, which considers the goal and the main tools for achieving it;
- The model of the green economy has been developed, unlike previous similar studies, it is hierarchical with isolated system-forming elements, with a clearly defined control element;
- The list of criteria for the green economy is developed;
- Recommendations have been developed to apply the green economy criteria in the Simulink simulation system, considering uncertainty conditions.

If we generalize the results on the whole, then we can say that, following the list developed, the primary tasks for the transition and development of the green economy have been solved.

Scope of the application criteria. The scope of the criteria list proposed covers two main areas: in practice and scientific research.

In practice, evaluation criteria are used in the methodology for forecasting green economy development. It is necessary to develop plans and government programs with an environmental assessment of the planned activities' impact on the environment following the national legislation. Similar actions are performed for programs developed for several industries: energy, industry, forestry, agriculture, fisheries, transport, urban and rural development planning, land use, and many others.

The forecasting method allows at the initial stage of the program implementation to estimate:

- The relevance of the plan, primarily to promote sustainable development and considering the impact on public health;
- Risks for the environment and public health;
- Influence on other plans and programs;

- Quantify statistically the magnitude of environmental impacts and the impact on the health of the population from the implementation of programs (e.g., the magnitude of the geographical area of impact or the number of people involved);
- The impact of the program on other projects and areas of activity in terms of funding;
- Influence on strategic or vulnerable areas, nature protection zones that have a recognized protected status.

In research and development activities, the criteria are also likely to find a wide application. The nearest future research topics form a generalized indicator of the green economy's effectiveness for optimizing the main parameters. Another promising direction for further research has already been identified above. It is an imitation modeling of the green economy processes, considering the conditions of uncertainty, though the range of tasks to be solved will be virtually unlimited.

The proposed methodology of regional green economy assessment can be used by the regional or federal government in the budgeting process and strategic planning process for the following needs:

- Evaluation of environmental and economic programs by green economy criteria;
- Optimizing financial policy for stimulation of the green economy;
- Environmental monitoring because of the green economy development;
- Developing state plans because of the green economy.

CONCLUSION

This paper has tried to extract the causal relationships of sustainable development components in the Middle East and North Africa by extracting development indicators and designing the Granger causality model. Meanwhile, sustainable development indicators were extracted by summarizing 92 social, economic, environmental, and institutional variables and then combined using principal component analysis. After designing the indicators, dynamic data panel models were used to extract the causal relationships between the branches. This study shows well that the economic development achieved at the expense of environmental degradation in this region has never improved the non-economic components of sustainable development. This is unrealistic, and the revenue model of the region's countries is based on the sale of crude oil, gas, and mineral resources or traditionally based on the agricultural sector. And has provided the means to limit and weaken patriarchy. A conclusion is that the increasing use of natural resources (including the extraction of hydrocarbon reserves and mines), deforestation, or the expansion of polluting activities in the region is due to the development of zones in the region. Decreased social development, increased economic development, and ultimately reduced institutional development. What has been shown in the results of this article is the pattern of governance in the region over the past years. Undoubtedly, predicting the future trend requires extensive and varied future research and studies. But looking at the development patterns of rentier systems such as our country, it can be acknowledged that the life of this failed development chain depends more than anything on its revenue model and the survival of oil revenues in most countries in the MENA region. The above causal chain will also continue its life.

REFERENCES

Anderson, T. W., & Hsiao, C. (1981). Estimation of Dynamic Models with Error Components. *Journal of the American Statistical Association*, *76*(375), 598–606. doi:10.1080/01621459.1981.10477691

Apergis, N. (2004). Inflation, output Growth, Volatility and Causality: Evidence from Panel Data and the G7 Countries. *Economics Letters*, *83*(2), 185–191. doi:10.1016/j.econlet.2003.11.006

Arellano, M., & Bond, S. (1991). Some Tests of Specification for Panel Data: Monte Carlo Evidence and an Application to Employment Equations, Review of Economic Studies. *Blackwell Publishing*, *58*(2), 277–297.

Atukeren, E. (2007). A causal analysis of the R&D interactions between the EU and the US. Global. *Economic Journal (London)*, *7*(4), 1850121.

Baltagi, B. H. (2021). *Econometric analysis of panel data*. Springer Nature. doi:10.1007/978-3-030-53953-5

Bhattarai, M., & Hammig, M. (2001). Institutions and the environmental Kuznets curve for deforestation: A crosscountry analysis for Latin America, Africa and Asia. *World Development*, *29*(6), 995–1010. doi:10.1016/S0305-750X(01)00019-5

Booth, D. E. (1995). Economic democracy as an environmental measure. *Ecological Economics*, *12*(3), 225–236. doi:10.1016/0921-8009(94)00046-X PMID:12292356

Chen, B., & Feng, Y. (1996). Some political determinants of economic growth: Theory and empirical implications. *European Journal of Political Economy*, *12*(4), 609–627. doi:10.1016/S0176-2680(96)00019-5

Chen, S., Saud, S., Bano, S., & Haseeb, A. (2019). The nexus between financial development, globalization, and environmental degradation: Fresh evidence from Central and Eastern European Countries. *Environmental Science and Pollution Research International*, *26*(24), 24733–24747. doi:10.100711356-019-05714-w PMID:31240660

Coondoo, D., & Dinda, S. (2002). Causality between income and emission: A country group-specific econometric analysis. *Ecological Economics*, *40*(3), 351–367. doi:10.1016/S0921-8009(01)00280-4

Costantini, V., & Martini, C. (2010). The causality between energy consumption and economic growth: A multi-sectoral analysis using non-stationary cointegrated panel data. *Energy Economics*, *32*(3), 591–603. doi:10.1016/j.eneco.2009.09.013

Culas, R. J. (2007). Deforestation and the environmental Kuznets curve: An institutional perspective. *Ecological Economics*, *61*(2-3), 429–437. doi:10.1016/j.ecolecon.2006.03.014

Desai, R. M., Olofsgård, A., & Yousef, T. M. (2005). Inflation and inequality: Does political structure matter? *Economics Letters*, *87*(1), 41–46. doi:10.1016/j.econlet.2004.08.012

Diao, X. D., Zeng, S. X., Tam, C. M., & Tam, V. W. (2009). EKC analysis for studying economic growth and environmental quality: A case study in China. *Journal of Cleaner Production*, *17*(5), 541–548. doi:10.1016/j.jclepro.2008.09.007

Dinda, S. (2004). Environmental Kuznets curve hypothesis: A survey. *Ecological Economics*, *49*(4), 431–455. doi:10.1016/j.ecolecon.2004.02.011

Farzin, Y. H., & Bond, C. A. (2006). Democracy and environmental quality. *Journal of Development Economics*, *81*(1), 213–235. doi:10.1016/j.jdeveco.2005.04.003

Frederking, L. C. (2002). Is there an endogenous relationship between culture and economic development? *Journal of Economic Behavior & Organization*, *48*(2), 105–126. doi:10.1016/S0167-2681(01)00228-1

Freedom House Organization. (2020). *Freedom House: a history*. FHO. http://www. freedomhouse. org/template. cfm

Hartwig, J. (2010). Is health capital formation good for long-term economic growth?–Panel Granger-causality evidence for OECD countries. *Journal of Macroeconomics*, *32*(1), 314–325. doi:10.1016/j.jmacro.2009.06.003

Heller, P., Harilal, K. N., & Chaudhuri, S. (2007). Building local democracy: Evaluating the impact of decentralization in Kerala, India. *World Development*, *35*(4), 626–648. doi:10.1016/j.worlddev.2006.07.001

Hoffmann, R., Lee, C. G., Ramasamy, B., & Yeung, M. (2005). FDI and pollution: A granger causality test using panel data. Journal of International Development. *The Journal of the Development Studies Association*, *17*(3), 311–317.

Holtz-Eakin, D., Newey, W., & Rosen, H. S. (1988). Estimating vector autoregressions with panel data. *Econometrica*, *56*(6), 1371–1395. doi:10.2307/1913103

Hsiao, C. (1982). Autoregressive modeling and causal ordering of economic variables. *Journal of Economic Dynamics & Control*, *4*, 243–259. doi:10.1016/0165-1889(82)90015-X

Hsiao, F. S., & Hsiao, M. C. W. (2006). FDI, exports, and GDP in East and Southeast Asia—Panel data versus time-series causality analyses. *Journal of Asian Economics*, *17*(6), 1082–1106. doi:10.1016/j.asieco.2006.09.011

Jolliffe, I. (2005). *Principal component analysis*. Encyclopedia of statistics in behavioral science.

Joseph, J. (2001). Sustainable development and democracy in the megacities. *Development in Practice*, *11*(2-3), 218–231. doi:10.1080/09614520120056360

Justesen, M. K. (2008). The effect of economic freedom on growth revisited: New evidence on causality from a panel of countries 1970–1999. *European Journal of Political Economy*, *24*(3), 642–660. doi:10.1016/j.ejpoleco.2008.06.003

Kaufmann, D., Kraay, A., & Mastruzzi, M. (2009). Governance matters VIII: Aggregate and individual governance indicators, 1996-2008. *World Bank Policy Research Working Paper*, 4978. doi:10.1596/1813-9450-4978

Kónya, L. (2006). Exports and growth: Granger causality analysis on OECD countries with a panel data approach. *Economic Modelling*, *23*(6), 978–992. doi:10.1016/j.econmod.2006.04.008

Lee, C. C., Chang, C. P., & Chen, P. F. (2008). Energy-income causality in OECD countries revisited: The key role of capital stock. *Energy Economics*, *30*(5), 2359–2373. doi:10.1016/j.eneco.2008.01.005

Magnani, E. (2001). The Environmental Kuznets Curve: Development path or policy result? *Environmental Modelling & Software*, *16*(2), 157–165. doi:10.1016/S1364-8152(00)00079-7

Menegat, R. (2002). Participatory democracy and sustainable development: Integratedurbanenvironmentalmanagementin Porto Alegre, Brazil. *Environment and Urbanization*, *14*(2), 181–206. doi:10.1177/095624780201400215

Miljkovic, D., & Rimal, A. (2008). The impact of socio-economic factors on political instability: A cross-country analysis. *Journal of Socio-Economics*, *37*(6), 2454–2463. doi:10.1016/j.socec.2008.04.007

Miyakoshi, T., & Tsukuda, Y. (2004). The causes of the long stagnation in Japan. *Applied Financial Economics*, *14*(2), 113–120. doi:10.1080/0960310042000176380

Narayan, P. K., Nielsen, I., & Smyth, R. (2008). Panel data, cointegration, causality and Wagner's law: Empirical evidence from Chinese provinces. *China Economic Review*, *19*(2), 297–307. doi:10.1016/j.chieco.2006.11.004

Newman, B. A., & Thomson, R. J. (1989). Economic growth and social development: A longitudinal analysis of causal priority. *World Development*, *17*(4), 461–471. doi:10.1016/0305-750X(89)90255-6

Norouzi, N. (2019). An overview on Water, Energy & Environment by 2030. *International Journal of Management Perspective*, *8*(2), 11–19.

Norouzi, N. (2020). Climate change impacts on the water flow to the reservoir of the Dez Dam basin. *Water Cycle*, *1*, 113–120. doi:10.1016/j.watcyc.2020.08.001

Norouzi, N., & Dehghani, M. A. (2020). A Backward Scenario Planning Overview of the Greenhouse Gas Emission in Iran by the End of the Sixth Progress Plan. *Current Environmental Management (Formerly: Current Environmental Engineering), 7*(1), 13-35.

Norouzi, N., & Fani, M. (2020). Black gold falls, black plague arise-An Opec crude oil price forecast using a gray prediction model. *Upstream Oil and Gas Technology*, *5*, 100015. doi:10.1016/j.upstre.2020.100015

Norouzi, N., & Fani, M. (2021a). The seventh line: A scenario planning strategic framework for Iranian 7th energy progress plan by 2020-2025. *Journal of Energy Management and Technology*, *5*(3), 43–53.

Norouzi, N., & Fani, M. (2021b). The prioritization and feasibility study over renewable technologies using fuzzy logic: A case study for Takestan plains. *Journal of Energy Management and Technology*, *5*(2), 12–22.

Norouzi, N., Fani, M., & Ziarani, Z. K. (2020). The fall of oil Age: A scenario planning approach over the last peak oil of human history by 2040. *Journal of Petroleum Science Engineering*, *188*, 106827. doi:10.1016/j.petrol.2019.106827

Norouzi, N., & Kalantari, G. (2020). The sun food-water-energy nexus governance model a case study for Iran. *Water-Energy Nexus*, *3*, 72–80. doi:10.1016/j.wen.2020.05.005

OECD. (2008). *Handbook on constructing composite indicators: methodology and user guide*. OECD.

Olson, M. (1993). Dictatorship, democracy, and development. *The American Political Science Review*, *87*(3), 567–576. doi:10.2307/2938736

Piñeiro Chousa, J., Tamazian, A., & Vadlamannati, K. C. (2017). Does higher economic and financial development lead to environmental degradation: Evidence from BRIC countries. *Energy Policy*, *37*(1), 2009.

Reporters without Borders. (2020). *Press Freedom Index*. RSF. www.rsf.org

Sachikonye, L. M. (2002). Democracy, Sustainable Development and Poverty: Are they Compatible? *Occasional Paper*, (2), 4. Development Policy Management Forum (DPMF).

Shahbaz, M., Haouas, I., Sohag, K., & Ozturk, I. (2020). The financial development-environmental degradation nexus in the United Arab Emirates: The importance of growth, globalization and structural breaks. *Environmental Science and Pollution Research International*, *27*(10), 1–15. doi:10.100711356-019-07085-8 PMID:31950417

Shahbaz, M., Nasir, M. A., & Roubaud, D. (2018). Environmental degradation in France: The effects of FDI, financial development, and energy innovations. *Energy Economics*, *74*, 843–857. doi:10.1016/j.eneco.2018.07.020

Söderbaum, P. (2007). Issues of paradigm, ideology and democracy in sustainability assessment. *Ecological Economics*, *60*(3), 613–626. doi:10.1016/j.ecolecon.2006.01.006

Stern, D. I. (2004). The rise and fall of the environmental Kuznets curve. *World Development*, *32*(8), 1419–1439. doi:10.1016/j.worlddev.2004.03.004

Stern, D. I., Common, M. S., & Barbier, E. B. (1996). Economic growth and environmental degradation: The environmental Kuznets curve and sustainable development. *World Development*, *24*(7), 1151–1160. doi:10.1016/0305-750X(96)00032-0

Tamazian, A., Chousa, J. P., & Vadlamannati, K. C. (2009). Does higher economic and financial development lead to environmental degradation: Evidence from BRIC countries. *Energy Policy*, *37*(1), 246–253. doi:10.1016/j.enpol.2008.08.025

ADDITIONAL READING

Adams, S., & Klobodu, E. K. M. (2018). Financial development and environmental degradation: Does political regime matter? *Journal of Cleaner Production*, *197*, 1472–1479. doi:10.1016/j.jclepro.2018.06.252

Chen, S., Saud, S., Bano, S., & Haseeb, A. (2019). The nexus between financial development, globalization, and environmental degradation: Fresh evidence from Central and Eastern European Countries. *Environmental Science and Pollution Research International*, *26*(24), 24733–24747. doi:10.100711356-019-05714-w PMID:31240660

Halkos, G. E., & Polemis, M. L. (2017). Does financial development affect environmental degradation? Evidence from the OECD countries. *Business Strategy and the Environment*, *26*(8), 1162–1180. doi:10.1002/bse.1976

Nasir, M. A., Huynh, T. L. D., & Tram, H. T. X. (2019). Role of financial development, economic growth & foreign direct investment in driving climate change: A case of emerging ASEAN. *Journal of Environmental Management*, *242*, 131–141. doi:10.1016/j.jenvman.2019.03.112 PMID:31029890

Norouzi, N., & Dehghani, M. A. (2020). A Backward Scenario Planning Overview of the Greenhouse Gas Emission in Iran by the End of the Sixth Progress Plan. *Current Environmental Management (Formerly: Current Environmental Engineering), 7*(1), 13-35.

Piñeiro Chousa, J., Tamazian, A., & Vadlamannati, K. C. (2017). Does higher economic and financial development lead to environmental degradation: Evidence from BRIC countries. *Energy Policy*, *37*(1), 2009.

Shahbaz, M., Haouas, I., Sohag, K., & Ozturk, I. (2020). The financial development-environmental degradation nexus in the United Arab Emirates: The importance of growth, globalization and structural breaks. *Environmental Science and Pollution Research International*, *27*(10), 1–15. doi:10.100711356-019-07085-8 PMID:31950417

Shahbaz, M., Nasir, M. A., & Roubaud, D. (2018). Environmental degradation in France: The effects of FDI, financial development, and energy innovations. *Energy Economics*, *74*, 843–857. doi:10.1016/j.eneco.2018.07.020

Yahaya, N. S., Mohd-Jali, M. R., & Raji, J. O. (2020). The role of financial development and corruption in environmental degradation of Sub-Saharan African countries. *Management of Environmental Quality*, *31*(4), 895–913. doi:10.1108/MEQ-09-2019-0190

KEY TERMS AND DEFINITIONS

Circularity: A circular economy (also referred to as "circularity") is an economic system that tackles global challenges like climate change, biodiversity loss, waste, and pollution. Most linear economy businesses take a natural resource and turn it into a product that is ultimately destined to become waste because it has been designed and made. This process is often summarised by "take, make, waste." By contrast, a circular economy uses reuse, sharing, repair, refurbishment, remanufacturing, and recycling to create a closed-loop system, minimize resource inputs, and create waste, pollution, and carbon emissions. The circular economy aims to keep products, materials, equipment, and infrastructure in use for longer, thus improving the productivity of these resources. Waste materials and energy should become input for other processes through waste valorization: either as a component or recovered resource for another industrial process or as regenerative resources for nature (e.g., compost). This regenerative approach contrasts with the traditional linear economy, which has a "take, make, dispose of" production model.

Eco Commerce: Eco commerce is a business, investment, and technology-development model that employs market-based solutions to balancing the world's energy needs and environmental integrity. Through green trading and green finance, eco-commerce promotes the further development of "clean technologies" such as wind power, solar power, biomass, and hydropower.

Eco-Tariffs: An Eco-tariff, also known as an environmental tariff, is a trade barrier erected to reduce pollution and improve the environment. These trade barriers may take the form of import or export taxes on products with a large carbon footprint or imported from countries with lax environmental regulations.

Emissions Trading: Emissions trading (also known as cap and trade, emissions trading scheme, or ETS) is a market-based approach to controlling pollution by providing economic incentives for reducing the emissions of pollutants.

Environmental Enterprise: An environmental enterprise is an environmentally friendly/compatible business. Specifically, an environmental enterprise is a business that produces value in the same manner which an ecosystem does, neither producing waste nor consuming unsustainable resources. In addition, an environmental enterprise rather finds alternative ways to produce one's products instead of taking advantage of animals for the sake of human profits. To be closer to being an environmentally friendly company, some environmental enterprises invest their money to develop or improve their technologies which are also environmentally friendly. In addition, environmental enterprises usually try to reduce global warming, so some companies use environmentally friendly materials to build their stores. They also set in environmentally friendly place regulations. All these efforts of the environmental enterprises can bring positive effects both for nature and people. The concept is rooted in the well-enumerated theories of natural capital, the eco-economy, and cradle-to-cradle design. Examples of environmental enterprises would be Seventh Generation, Inc., and Whole Foods.

Green Economy: A green economy is an economy that aims at reducing environmental risks and ecological scarcities and that aims for sustainable development without degrading the environment. It is closely related to ecological economics but has a more politically applied focus. The 2011 UNEP Green Economy Report argues "that to be green, and an economy must be not only efficient but also fair. Fairness implies recognizing global and country-level equity dimensions, particularly in assuring a Just Transition to an economy that is low-carbon, resource-efficient, and socially inclusive."

Green Politics: Green politics, or ecopolitics, is a political ideology that aims to foster an ecologically sustainable society often, but not always, rooted in environmentalism, nonviolence, social justice, and grassroots democracy. It began taking shape in the western world in the 1970s; since then, Green parties have developed and established themselves in many countries around the globe and have achieved some electoral success.

Low-Carbon Economy: A low-carbon economy (LCE) or decarbonized economy is based on low-carbon power sources with minimal greenhouse gas (GHG) emissions into the atmosphere, specifically carbon dioxide. GHG emissions due to anthropogenic (human) activity are the dominant cause of observed climate change since the mid-20th century. Continued emission of greenhouse gases may cause long-lasting changes worldwide, increasing the likelihood of severe, pervasive, and irreversible effects for people and ecosystems.

Natural Resource Economics: Natural resource economics deals with the supply, demand, and allocation of the Earth's natural resources. One main objective of natural resource economics is to understand better the role of natural resources in the economy to develop more sustainable methods of managing those resources to ensure their future generations. Resource economists study interactions between economic and natural systems intending to develop a sustainable and efficient economy.

Sustainable Development: Sustainable development is an organizing principle for meeting human development goals while simultaneously sustaining the ability of natural systems to provide the natural resources and ecosystem services on which the economy and society depend. The desired result is a state of society where living conditions and resources are used to continue to meet human needs without undermining the integrity and stability of the natural system. Sustainable development can be defined as development that meets the needs of the present without compromising the ability of future generations to meet their own needs. Sustainability goals, such as the current UN-level Sustainable Development Goals, address the global challenges, including poverty, inequality, climate change, environmental degradation, peace, and justice.

Chapter 6
Peer-to-Peer Energy Trading Using Blockchain in Sub-Saharan Africa:
Towards a Policy and Regulatory Framework

Mirana Njakatiana Andriarisoa
Institute for Water and Energy Sciences Including Climate Change, Pan African University, Germany

David Tsuanyo
National Committee for Development of technologies, Ministry of Scientific Research and Innovation, Cameroon

Erick G. Tambo
Institute for Environment and Human Security, United Nations University, Germany

Axel Nguedia Nguedoung
Institute for Environment and Human Security, United Nations University, Germany

ABSTRACT

Peer-to-peer (P2P) energy trading using blockchain is presented as a great innovative potential to promote rural electrification. Opportunities and challenges assessment for the implementation of this technology in Sub-Saharan Africa shows that it is only at its embryonic stage in the region. The decreasing cost of stand-alone solar technology and the expansion of investment in mini-grid sector are among the opportunities. However, the considerable restriction of private participation in the mini-grid sector, the difficulty of the regulatory process and licensing requirements, the issues with tariff framework, and the uncertainty of the regulation about the future grid integration are among the main challenges. This chapter proposes a policy and regulation framework for the promotion of P2P energy trading using blockchain in Sub-Saharan Africa.

DOI: 10.4018/978-1-7998-8638-9.ch006

INTRODUCTION

The implementation of energy systems in Africa is facing many challenges where one of the most important lies in the financial viability of the project although the technical reliability has been proven (Cabanero et al., 2020). Over the past few decades, digital technologies and innovation have been continuously affecting the energy sector and led to new business models, both for national utility grids and for decentralized systems (off-grids, mini-grids). The progress in digital technologies has allowed the promotion of effective management of the energy demand and therefore makes energy systems more economical and environmentally friendly (Majeed Butt et al., 2021).

At the heart of this innovative energy management system is the peer-to-peer (P2P) energy sharing concept which is a decentralized electricity trade between prosumers and consumers (Alladi et al., 2019). Trading based on P2P models makes renewable energy more accessible, empowers consumers, and allows them to make better use of their energy resources (IRENA, 2020). In recent years, P2P trading platforms have been developed considerably worldwide. Many projects have already been implemented such as the Brooklyn Microgrid in the United States, the Centrica plc in the United Kingdom, the Lumenaza in Germany, and the SolShare in Bangladesh. Considerable benefits for the energy management of the communities were implemented as a result of those projects. Therefore, promoting P2P energy trading can be very promising for electrification within Sub-Saharan Africa (SSA) where 548 million people, which is 47% of the population, in the region lack access to electricity in 2018 (IEA et al., 2020).

In addition, the emergence of new digital technologies like artificial intelligence, the internet of things, and particularly blockchains can have a very promising considerable impact on energy access finance since it can lead to the creation of new innovative marketplaces more transparent, secured, and tamper-resistant (Ndung'u & Signé, 2020). In recent years, different applications of blockchain have been continuously developed. More than one-third of the use of blockchain in the energy sector is for P2P energy trading (IRENA, 2019).

However, P2P energy trading using blockchain is just at the initial stages in Africa either in research or in project implementation. Only a few companies are currently exploring that technology in the region. For instance, there is Rehub in Kenya currently having a pilot project. There is also the Lightency exploring the use of this technology in Tanzania, Uganda, Burkina Faso, Senegal, Mali, and Lybia where the policy is more favorable according to the company.

Nevertheless, the implementation of these new technologies in SSA is still limited despite their vast potential. One of the major barriers to its implementation is the lack of an adequate policy and regulatory framework (Klein et al., 2019). Policies are a prerequisite for other actions to follow. The current energy regulatory schemes all over the world and particularly in SSA were established based on the traditional energy market system which is centralized, vertical, and unidirectional. Hence, the regulatory framework for the case of revolutionized energy market such as decentralized and distributed energy market are not clear, even unavailable. Besides, technology is changing, thus it requires dynamic thinking (Blimpo & Cosgrove-Davies, 2019). Jordan et al. established a prospective architecture for P2P energy sharing in a South African context in 2018 and stated that the local grid needs to be a smart grid and deregulated in order to implement P2P energy trading. Deregulation will allow consumers to purchase electricity from prosumers, or from the grid (Jordan et al., 2018). The latest SDG 7 tracker report stated the importance of updating the policy framework to embrace and support innovation such as off-grid solutions and innovative business models (IEA et al., 2020). This chapter firstly carries out a review and analysis of the current situation on digital technology and blockchain situation in Sub-Saharan Africa and identifies the challenges and opportunities related to the development of this technology in the region, and secondly, proposes conceptual modeling of the policy and regulation framework necessary for the implementation of P2P energy trading using blockchain in SSA. The findings carried out in this chapter will help energy policymakers in Africa to acknowledge the importance of digital technology for the energy sector and to include that point in their energy policy strategies.

A descriptive-qualitative method was used to conduct the research. It consists of a detailed description to define the current situation of policy and regulation concerning P2P energy trading and blockchain technology within SSA based on the analysis of a company implementing the technology. Data collection methods included the application of documentary and literature review from desktop research to gather secondary data for the assessment of the digital technology situation in SSA and data on the use of advanced ICT technology such as Blockchain in those countries. It is also to identify what are the opportunities as well as the challenges in that field. The data collected was mainly generated from secondary literature such as reports from ministries of energy, the ministries of ICT (Information and Communications Technology), and international organizations specialized in digital technologies like the International Telecommunications Unions (ITU). Then, an analysis of the activities and processes of a company implementing P2P technologies was concluded along with an interview to collect the primary data necessary for a practical understanding of the reality on P2P energy trading and blockchain implementation within Africa, to know the challenges that the innovators are facing and to see their perspectives regarding the policy and regulations. The interview was conducted with the Rehub company which is currently conducting a pilot study on P2P energy trading using blockchain in Kenya.

BACKGROUND

Use of Blockchain in the Energy Sector and the Mechanism of P2P Energy Trading

Some research such as the PwC Global Fintech Report in 2016 and the IRENA innovation landscape brief on blockchain have explained the opportunities presented by blockchain for the energy sector (PricewaterhouseCoopers, 2016) and (IRENA, 2019). The different key applications integrating renewables can be summarized in figure 2 where Peer-to-peer energy trading is the most common application.

Figure 1. Blockchain-based key energy application in the renewable energy sector (IRENA, 2019)

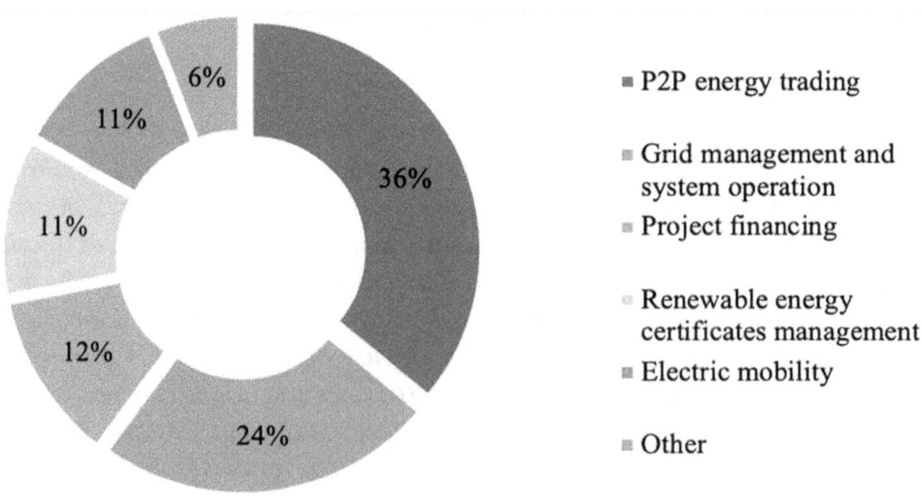

Peer-to-peer energy trading is the most common use of blockchain in the energy sector globally since it occupies 36% of the use of blockchain. After the feed-in tariff, Peer-to-peer energy trading is the next generation energy management technique for the smart grid. Peer-to-peer energy trading is a bottom-up energy market structure where the prosumers are connected directly with each other. It is a decentralized, autonomous, and flexible energy trading network. In energy trading, compared to existing FiT (Feed-in Tariff) schemes, the direct involvement of the users with one another and with the grid makes P2P systems unique (Tushar et al., 2018). A peer-to-peer energy network is a distributed network architecture in which the participants share a part of their energy resources with one another. Tushar et al. show that a P2P network can be divided into two layers: first, a virtual layer comprised of the information system, the market operation, the pricing mechanism, and the energy management system and the second one is the physical layer which is mainly the physical network that shares the electricity from the producers or sellers to the buyers after the financial transaction are settle in the virtual layer (Tushar et al., 2020).

Chenghua Zhang et al. (2018) performed a study on peer-to-peer energy trading in Microgrid in which they demonstrated that not only do direct transaction practice leads to the promotion of the use of renewable energy and reduction of transmission loss, but also it can generate revenues for producers and prosumers and save money for end users. P2P energy trading also constitutes an incentive for investment in renewable energy and ensures the balance of supply and demand.

Blockchain is considered a very promising technology for energy distribution and energy trading. Wongthongtham et al. (2021) have explored the blockchain technology for P2P energy trading and its implications given the trilemma: scalability, security, and decentralization. The general focus of the approach of blockchain is to provide a replicable data structure, sharable among the participants allowing secured, transparent, and decentralized energy trading in the P2P network. The popular methods for that are smart contract, Elecbay, consortium blockchain, Hyperledger, and Ethereum (Alladi et al., 2019).

Blockchain Features for P2P Energy Trading and its Benefits for SSA

Introducing blockchain in the mini-grid sector in SSA represents several promising benefits for the sector. Based on the distributed energy systems, the electric mini-grid transactions require a trusty system of payment. Sana

Noor et al. (2018) demonstrated that the emerging blockchain technology can ensure an efficient and robust trading system with low cost through the implementation of optimal energy management strategies.

Blockchain technologies are introduced to facilitate peer-to-peer transactions to enhance the optimization of supply and demand management. Therefore, they suggested that an extended Demand Side Management (DSM) framework and optimized strategy would help consumers reduce their electricity bills, and would allow them to manage their own load by scheduling their electricity consumption. Back in 2010, The Association for Computing Machinery (ACM) shared a view on cloud computing and developed a testing software project for blockchain. 23 utility companies have participated in the project testing and attested the success of blockchain in their work and it pays off when used for trading (Armbrust et al., 2010).

Noor et al. demonstrated as well that exploring the potential of blockchain would reduce the stress on the grid and avoid building new generation capacity (Noor et al., 2018). The use of blockchain is a great way to avoid passing through national utilities or central electricity agency that often fails in managing efficiently the supply and demand and who are highly affected by corruption issue. Münsing et al. (2017) presented similar research that blockchain is a great solution for several problems of optimization

In addition, blockchain use in the distributed energy system would improve reliability by allowing real-time control of the power supply and flow (Mylrea et al., 2017). That leads to the building of a resilient system where the participants trust each other and that would improve considerably the data management. Contrary to the vulnerable traditional system where data is stored in one point, data is stored across all nodes in the distributed renewable energy enhanced by blockchain, which makes it easier to track and audit (Khatoon et al., 2019)

Also, promoting blockchain-enabled contracts would allow transparent energy transactions. In 2017, Esther Mengelkamp and her colleagues also shared a study of a micro-grid energy market for the case study of micro-grid Brooklyn where prosumers can sell and buy self-produced energy (Mengelkamp et al., 2017). The Brooklyn microgrid market was the first project in the world that facilitated a peer-to-peer energy transaction using blockchain. Hence, the utilization of blockchain technology will improve trade and communication among prosumers by increasing transparency of transactions along with fortifying security due to a universal decentralized ledger system.

Samuel et al. have stated that Blockchain is particularly suitable for rural settlers, especially for electrification in SSA where the households owning a solar home system (SHS) can generate electricity and trade their surplus among participants in the same community (Samuel et al., 2020). When the SHSs generate enough energy, which can meet consumers' demands, they get incentives instantly. Energy trading is operated in the blockchain network through digital tokens and tokens can be redeemed for a remote cryptocurrency Blockchain technology improves the pricing mechanism of P2P energy trading by removing any third parties. Besides, they have performed a comparative study in Sub-Saharan Africa on the energy cost for the conventional scenarios not using Blockchain and a future scenario integrating blockchain and concluded that the future scenario with blockchain can provide further cost reductions via incentives and P2P energy trading. They stated that the advantage of cryptocurrency trading is that there are clear working capital costs and electricity bill reduction targets. Their analysis shows that while using Blockchain, the Levelized Cost of Energy (LCOE) presents a 95% decline in the cost of battery storage and a 75% fall in the solar module. Furthermore, the anticipated future LCOE represents less spatial variability among different locations because while the spatial variation was 0.15$/kWh in the current system, it will decrease to 0.049$/KWh when using Blockchain (Samuel et al., 2020).

However, peer-to-peer energy trading is just at a very early stage in Africa either in a research study or in project implementation. That is why the present research would focus on blockchain use in P2P energy trading and is aiming at analyzing the policy and regulation required to promote the establishment of this practice in the region.

Analysis of Successful Cases of P2P Energy Trading

Several P2P energy trading projects have been developed around the world. This section presents the cases of some companies in the United States, Europe, Australia, and Asia, and then provides a brief analysis of the opportunity.

- **Grid+** is a company based in Texas which provides simplified financial security. It is a supplier of distributed electric power (Consensys, 2018). The company is proving the concept of using blockchain technology to revolutionize traditional methodologies and it uses an already established blockchain infrastructure to facilitate efficient electric power trading within the market. The cryptocurrency used by Grid plus is Ethereum which will automate its energy billing system and provide transparency by enabling consumers to access their energy usage data in real-time via an in-home smart agent device. Grid+ is experiencing exponential growth from 1,000 households to 255,000 households in just over a year and has demonstrated early progress in product development. However, peer-to-peer energy trading in the United States encounters challenges due to regulatory restrictions. In 2019, GridPlus started extending its project to Singapore which is a highly suitable early market because it is liberalized, stable, and has well-defined regulations. Besides, the government is very favorable of not only the technology but also the regulation, providing incentives to innovation.

- **Power Ledger** is an Australian renewable energy and environmental commodities trading company which allow its participants to invest in renewable and transact energy. They have created a software that enables peer-to-peer energy trading from solar photovoltaics on rooftops. Through the use of blockchain technology, they empower households to sell to the neighborhood their excess power. They also aim at creating new markets for energy from renewable sources. Participants can use or sell the power produced through their own device and the revenue can be automatically allocated to their personal wallet address proportionally to their shareholding ratio. Working with Thai Digital Energy Development (TDED), the Power Ledger utilized blockchain technology to manage several clean energy projects. Power Ledger also has a partnership with Malaysia's Sustainable Energy Development Authority to help increase the country's renewable energy generation to 20% of the total generated. Power Ledger has saved an average of USD 424 on annual electricity bills for its energy consumers and helped solar home system owners double the savings they normally get from their solar plants (Kabessa, 2017). However, the Australian P2P energy trading is also facing a challenge regarding regulation, since regulations to enable and assess the P2P power trading market did not exist yet. Also, considering the current Australian regulation, it is challenging to implement a low-cost P2P administration (ARENA, 2017)
- **LO3 Energy** is a blockchain-based community energy platform headquartered in New York. It has developed an energy platform that enhances the integration of distributed energy resources (DERS) in Brooklyn. The platform promotes the use of renewables like solar and wind and includes battery storage with supply networks on the grid. The company is teaming up with other partners such as Siemens,

Centrica, and Braemar Energy Ventures. Last July, Shell Ventures and Sumitomo also joined the business. They are also undertaking similar projects in Colombia, Australia, the UK, and Japan (Gordon, 2019).

- **The Brooklyn microgrid project** has demonstrated that a private blockchain protocol can successfully operate a microgrid energy market. The microgrid setup, the grid connection, and the information system are fully operational. Then the market and pricing mechanism are partly implemented but still need to be tested and adequately adapted for allocation efficiency. Finally, the implemented Energy management trading systems are also operational but require further development. Besides, public acceptance and customer participation are very promising for the case of the Brooklyn microgrid project. However, the legal environment did not allow peer-to-peer electricity transactions between participants. Therefore, there is a need to adapt the regulation to fully develop that system (Mengelkamp et al., 2017).
- **ME SOLshare** company created in 2014 is based in Bangladesh and operated the first ever P2P energy exchange platform. It has developed a peer-to-peer platform that allows peer-to-peer trading between prosumers (Solshare, 2020). The benefit of solar energy is showcased by over 6 million households purchasing this form of the solar energy system. SOL share stated that all systems in Bangladesh are standard due to heavy regulations and the standard systems produce 30% excess electricity. With this logic having 10 solar-producing households can supply the electrical needs of 3 households without electricity. Swarm electrification exploits the excess electricity generated from each solar system. Bangladesh's current electricity cost is $1.50 KWh, while Solshare prices generation at $1.00 with a generation fee of 25%. ME SOLshare showed that using microgrids can help people out of poverty, with every $1 invested in SOLshare technology $4.85 was generated on social return of investment. Solshare expects to operate more than 20,000 nanogrids by the end of 2021, which are expected to supply more than 1,000,000 customers in Bangladesh. The success of Solshare projects lies first in the massive number of solar home systems in Bangladesh. The case of Bangladesh has demonstrated that transactive energy does not have to be complex and the technology enabling it can be cheap. Also, the presence of an advanced developing market with low regulation in Bangladesh represents an ideal environment to harness the technical and commercial potential for microgrids.
- **Suncontract:** SunContract is a blockchain-based company that is pioneering the decentralization of the energy sector. It launched the

> world's first trading platform in 2018 in Slovenia. The company has partnered with European nations and many energy sectors and blockchain partnerships to provide an energy trading platform to households. Via an app, users can enter into deals with each other, set prices, and share energy. While the project is currently being implemented in Slovenia, it will shortly begin in the European Union where they have received support from the government and reputable EU commissioners. The SunContract platform has allowed users to instantly access and audit energy consumption and production. It also ensures transparency and removes the need for intermediaries. In its white paper in 2017, Suncontract pointed out the issue of subsiding conventional energy in many markets and consumer segments and emphasizes the importance of a conducive policy framework as a prerequisite for clean energy. It highlighted the importance of the democratization of energy supply as well (SunContract, 2017). Regarding the legal context, contrary to the US, the European Union states are more favorable to peer-to-peer energy trading.

Generally, peer-to-peer energy trading always comes with the promises of lower electricity prices and more stable grids. That is the case for instance for Grid+. Nevertheless, in terms of real-world activity, operating a P2P energy trading is still challenging in many cases, especially in the U.S. which has restrictions in terms of regulation. On the contrary, the current state of European Union energy law might in principle allow P2P electricity trading, yet the lack of specific provisions leads to challenges in practice. According to Henri Van Soest (2019) who conduct a review of the legal context of P2P electricity trading, the electricity system in the US is far behind the European Union in terms of modern approaches to energy regulation. That is why most P2P energy trading projects are localized in Europe where regulatory efforts to initiate and promote P2P energy trading are carried out. In 2018, the European Union mandated that all their member states should facilitate the study and implementation of projects in P2P energy trading by 2021 (Deing, 2019).

Importance of Policy and Regulation Framework in P2P Energy Trading

The policy and regulatory framework are important requirements for the implementation of P2P energy trading. The lack of those regulations is one of the major barriers to the implementation of P2P energy sharing across the world (Klein et al., 2019). Efforts are being made to recognize the importance of regulatory schemes in facilitating the development of P2P energy trading and to deal with the related challenges (Ahl et al., 2019) and (Diestelmeier, 2019).

Mengelkamp et al. (2017) conducted a study on designing microgrid energy markets and had the Brooklyn Microgrid as a case study. They asserted that a microgrid market has seven components where the main one is the regulation which ensures that the microgrid energy system is in line with the current energy policy. The regulation and legal environment govern all the six other components of the micro-grid market which are: a micro-grid setup, the grid connections, a high-performing information system, the market mechanism, the pricing mechanism, and the energy management trading system (Mengelkamp et al., 2017). Tushar et al. (2018) also confirmed that effective policy and regulation are necessary to govern the success of P2P energy trading. Quite similarly to Mengelkamp's study, they analyzed that the required policy consists in deciding the kind of market design which will be allowed, the distribution of taxes and fees, and also determining how the status and mechanism of the P2P market will be in relation to the already existing energy market. However, they emphasize more on the role of governments. Since P2P energy transactions can accelerate renewable energy deployment and limit environmental issues, governments can decide to support P2P energy trading by establishing the right regulatory system, like in the EU countries. On the contrary, governments also have the power to discourage the implementation of a project if it is deemed to have considerable negative impacts on the existing energy system. Furthermore, the OECD (2019) added that the regulatory challenges include also smart contract validation, recognizing cryptographic signatures, and automating settlements. It is also necessary to deal with eventual barriers for communities to share their resources. Zhou et al. (2020) made a recapitulation of the elements that needed to be considered under the policy analysis of P2P energy trading which are: first, the DSO regulation, second the legitimacy and distribution of taxes and fees, third is the ownership and partnership models, prosumer licensing, market roles and the last is the decentralized responsibility, flexibility incentives, and customer protection. All those previous researches were conducted in developed countries' contexts. They can serve as basic studies for a similar study in Africa but it is also necessary to consider the context of the region, hence the present study which will assess the existing policy and context in Africa. For the specific case of SSA, hardly any research has addressed the question of P2P energy trading and the few existing pieces of research are focused more on the technical study, not on the policy aspect. Among them, Ma et al. (2018) researched P2P trading solutions for micro-grids in Kenya.

Although their study focused more on the technical aspect, it gave an overview of the influential factors to the microgrids in Kenya which are the climate, social culture, technology, economy, and policy and regulation. On this point, they specified that in 2014, Kenya allowed a private company to build and operate microgrids for the first time while being open to alternative options for grid expansion. However, an in-depth policy and regulation analysis was not completed.

Diesteleimer (2019) focused more on the policy implications of the new roles of consumers for EU electricity law. One of the pressing legal questions of the energy transition is how to integrate prosumers into the electricity market since they have limited roles thus far. Due to blockchain technology, the approach of "integration in the market" can be changed to "becoming the market". Three main policy implications of blockchain-based electricity transactions were identified, which are: organizing decentralized responsibilities, establishing a legal framework for incentivizing consumers to invest in flexible technologies, and the last is to strike balance between self-responsibility and protection of consumers. Campos Inês et al. (2020) also focused their study on the regulatory challenges and opportunities for collective renewable energy prosumers in the EU and pointed out that those types of prosumers may require a specific legal framework such as the possibility to sell electricity surplus directly to another consumer. Apart from that, Morstyn et al. (2018) highlighted the role of P2P energy trading markets for distribution system operators to facilitate prosumer coordination. Therefore, they suggested that changes in Distribution System Operators (DSO) regulations should be made to link the DSO rate of return to network capacity investments to provide incentives for DSOs. They also suggested carrying out further investigations to understand how regulations can best align DSO, prosumer, social, and power system objectives. Morstyn's work contributes considerably to P2P trading and DSOs but does not integrate much the concept of the blockchain (Morstyn et al., 2018). Still on the role of consumers, according to the global blockchain policy forum in 2019, with the decentralization of the energy system, investment is shifting from utilities to consumers or third parties. Therefore, new challenges will emerge for the energy sector, for instance reinforcing consumer laws and socializing properly the cost of the network infrastructure. The role of prosumers in the electricity trade is at the heart of the regulatory challenges for P2P trading (OECD, 2019). However, in Africa, the concept of the prosumer is just at its infant stage since the energy market is still more into the traditional system of centralized energy supply. In companies exploring the approach of P2P in SSA, the new roles of consumers are not currently fully defined and there is no appropriate regulation addressing that issue. Therefore, this research will be necessary to highlight the policy and regulations necessary to embrace the new roles of customers.

When integrating blockchain in the P2P energy trading context, Ahl et al. (2019) conducted an analytical framework for blockchain-based P2P microgrids which includes technological, economic, social, environmental, and institutional dimensions. In the institutions, it affects the market policy, grid codes, P2P policy, and mechanism for institutional innovation. Their research contributes to the exploration of potential challenges of blockchain-based P2P microgrids and proposes practical implications for institutional development which can be applied to the context of Africa where institutions do not play enough roles in promoting new technologies. However, this research is too broad since it embraces different dimensions, but did not develop and go deep into the policy and regulation aspects, so the present work consists in focusing mainly on the institutions' involvement in the implementation of blockchain-based P2P energy trading.

Also, policy should be constantly reviewed to grow with and encourage innovation within the blockchain industry. Governments' collaboration with innovators is crucial to implement mechanisms like regulatory sandboxes and consider international approaches. The connection between the government and innovators is also necessary to define and establish the right regulatory framework. (OECD, 2019). This government collaboration is still lacking in Africa and the present research is aiming partly at identifying the actions required from African governments to promote the technology in the region.

Generally, P2P energy trading always comes with the promises of lower electricity prices and more stable grids. Nevertheless, in terms of real-world activity, operating a P2P energy trading is still challenging in many cases, especially in the U.S. which has restrictions in terms of regulation. Conversely, the current state of European Union energy law might in principle allow P2P electricity trading, yet the lack of specific provisions leads to challenges in practice. According to Henri Van Soest (2019) who conduct a review of the legal context of P2P electricity trading, the electricity system in the US is far behind the European Union in terms of modern approaches to energy regulation. That is why most of the P2P energy trading projects are localized in Europe that carried out regulatory efforts to initiate and promote P2P energy trading. In 2018, the European Union mandated that all their member states should facilitate the study and implementation of projects in P2P energy trading by 2021 (Deing, 2019). In many other situations around the world, P2P energy trading is explicitly forbidden due to the fervent lobbying power of incumbents and the energy market which is less liberalized. Those studies show that although P2P energy trading and blockchain are already more advanced in American and European countries, the legal environment is still not properly well defined yet. Therefore, for the African cases where P2P technology is just at an early stage, it would be interesting to study the legal environment before fully promoting the technology to avoid confusing or ambiguous situations while the technology is already implemented. Therefore, current research contributions in assessing the legal environment within SSA are limited.

To sum up, several gaps were identified in the existing research. First, most of the existing research on P2P energy trading was mainly conducted in developed countries and does not consider the contexts in developing regions such as SSA. The present research is among the initial set of research dedicated to exploring the implementation of P2P energy trading in underdeveloped countries that one would find within SSA. Secondly, most of the existing research did not develop and go deep into the policy and regulation aspect. Instead, they are mainly focused on the technical aspect of the market design and only an overview was presented on the policy and regulation aspect. Although several studies have discussed the P2P energy trading policy and blockchain from different perspectives, there is still a lack of details on how the new policy reform should be exactly. Therefore, the present research aims to bring contribution to the policy analysis of P2P energy trading and blockchain uses.

OPPORTUNITIES AND CHALLENGES FOR DEVELOPING P2P ENERGY TRADING USING BLOCKCHAIN IN SUB-SAHARAN AFRICA

Africa's ICT Development Indicator and Blockchain Situation in Sub-Saharan Africa

- **A state analysis of ICT and blockchain in SSA**

Rural Sub-Saharan Africa is still largely under-electrified. Mini-grids, micro-grids, and solar home systems have been promoted in recent years to provide modern energy services as well as new sources of employment to remote communities, and that is facilitated by digital technologies and payment tools (IEA, 2019). Nevertheless, the African Economic Outlook 2020 stated that African countries are digitally under-connected and lag behind in terms of the digital revolution (AfDB, 2020) although the continent has witnessed considerable growth in ICT access since the 1990s. The development in terms of ICT in Africa is predominantly driven by mobile telecommunication and particularly mobile financial services such as M-PESA in Kenya. In fact, in 2018, Africa alone disposes of half of the global mobile money accounts and it will continue growing fast until 2025. Also, internet penetration is one of the most used ICT development indicators and although it is very low in Africa compared to the rest of the world, it has considerable increase over the last few years (Kapoor et al., 2019)

Furthermore, the continent is starting to be attracted by the 4th industrial revolution. Njuguna Ndung'u and Landry Signé recognized that "The Fourth Industrial Revolution and digitization will transform Africa into a global powerhouse". In recent years, artificial intelligence and blockchain have started being attractive to the continent (Ndung'u & Signé, 2020). Kapoor et al.'s survey on African firms demonstrated that most of the African firms reported moderate to very low levels of business preparedness for the Blockchain, artificial intelligence, and the internet of things. However, particularly for blockchain, although more than 10% have very low awareness about it, more than 20% are highly aware of the technology (Kapoor et al., 2019).

Most SSA countries have not started to analyze the potential that blockchain technology can generate and do not have any policies or regulations related to that matter. However, some see blockchain as an uncontrollable tool that can threaten the financial status of the country and has set restrictive policies in a bid to restrict this technology. For instance, according to Mc Kenzie's research, blockchain and cryptocurrency are banned or prohibited in Zambia, Zimbabwe, Namibia, and Swaziland. On the contrary, other countries have set friendly and progressive policies toward blockchains, such as South Africa, Mauritius, Senegal, and Sierra Leone. Some others are indifferent or do not have any official stance toward blockchain uses such as Kenya, Tanzania, Ethiopia, Madagascar, Uganda, Democratic Republic of Congo, Botswana, Cameroon, Nigeria, and Ghana (McKenzie, 2018). Ferguson also confirmed that most of the regulators in Africa were taking a "wait and see" approach to digital currency. However, South Africa and Kenya take "use-at-your own risk approach" (Ferguson et al., 2019).

Regarding cryptocurrency associated with blockchain technology, it is also at an early stage in Africa. The few available pieces of research in that field in Africa are reflecting that most African governments are still on stand-by towards that innovative finance. According to research by Pan African Ecobank conducted in 39 SSA countries, Africa's regulators and central banks are adopting a "Wait and See" approach. They remain reticent towards authorizing cryptocurrency transactions due to the potential risks that it may create. (Ecobank, 2018). Nevertheless, an article written by Pavitra Rao in the Africa Renewal in 2018 stated that the conditions in Africa are favorable for virtual currency and Africa could be the next frontier for cryptocurrency (Rao, 2018). Bitcoin is the most common cryptocurrency used in the continent and Botswana, Ghana, Kenya, Nigeria, South Africa, and Zimbabwe are the main Bitcoin countries in Africa. Besides, research conducted by T. Koffman (2019) confirmed also that the blockchain situation in Africa is raising.

Cryptocurrency has started popping up for remittance services in the continent for instance Abra in Malawi and Morocco, BitMari in Zimbabwe, GeoPay In South Africa, and Kobocoin in Nigeria. Particularly in Kenya, the BitPesa facility was launched in 2013. It offers virtual remittances to African and international countries, to and from individuals' mobile wallets. According to LocalBitcoins.com, the value of bitcoin trading exceeded already 1.8 million USD as of December 2017 (BitPesa, 2020). In Senegal, the digital currency eCFA has been introduced in December 2016 (Rao, 2018).

In South Africa, in 2015, A. Nieman performed a research on virtual currency entitled "A few South African cents' worth on bitcoin". It has developed an assessment of the current state of the development of bitcoin use in South Africa. It also analyzed the challenges that this new technology is facing. Particularly, in April 2015, South Africa held its first commercial bitcoin conference in Cape Town. There are local innovators like Lorian Gamaroff in the country, who designed a local blockchain smart metering service for energy management and prepaid electricity. In addition, the Cape Innovation and Technology Initiative has launched a Virtual Currency hub and incubator. Since 2012, regular Bitcoin meetups are organized in Capetown and Stellenbosch (Nieman, 2015).

In Malawi, up to 10,000 farmers joined the one-year pilot project intended to explore the use of blockchain technology to track supply chains for tea sold to the great companies Unilever and the British supermarket Sainsbury's. The project consisted in rewarding with financial incentives like access to credit and preferential loans those producing a fairer and more sustainable brew. To a further extent, a group of ten large food and retail companies (Nestle, Unilever, Tyson Foods, etc.) has joined the IBM project on how blockchain systems can help track the food supply chain and improve safety (BNP Paribas, 2019).

There is also the Plaas farmers token, which is a technology used by the agricultural business plaasio founded in Gaborone Botswana in 2017. It is working as a mobile app enabling farmers to manage their stock on blockchain. It is a platform designed for Africans working in the agriculture sector for data storage, tracking, and validation with price notification via SMS and market improvement via future contracts (PLAAS, 2018).

- **The current use of blockchain in the energy sector in SSA**

Several African companies have started having an interest in exploring blockchain technology in the energy sector. For instance, there is the OneWattSolar based in Nigeria whose objective is to contribute to solving the unreliable power grid in Nigeria where many households and businesses are obliged to use diesel power generator which is expensive, noisy, and harmful to the environment. OneWattSolar is an innovative funding model. It pays for, installs, owns, and operates the Solar Home System with zero up-front investment and no out-of-pocket expenses from the households. The operational data is put onto Blockchain and tokens are used to pay for the electricity which ensures high efficiency and transparency for consumers, developers, investors, and regulators. OneWattSolar currently has almost 7000 customers with an electricity generation capacity of 4018 MWh. The tariff is very low compared to the one of the grid cause it is only 0.1USD cents per kWh (OneWattSolar, 2020).

There is also the South Africa-based startup Bankymoon which is developing smart meters with integrated payments using Bitcoin. They are using PAYG to top up the energy meters. Each meter has a unique bitcoin (Bankymoon, 2020).

Another company that is exploring blockchain use in Africa is Lightency. It is a green tech startup that harnesses the power of deep technologies to ensure better access to affordable and green energy by promoting distributed energy resources in microgrids. Particularly, they are exploring peer-to-peer trading platforms to balance the grid and allow the user to trade their energy surplus. The Lightency is targeting the following countries: Kenya, Tanzania, Uganda, Burkina Faso, Senegal, Mali, and Lybia. They claim that the government policies in those countries are favorable to peer-to-peer energy trading because, under a certain capacity of production, electricity can be traded freely between participants. Also, according to their study, the mini-grid market is already developed in those countries, which can promise wider customers for them. It is also because of the ease of doing business and some encouraging indicators such as the high mobile penetration (Lightency, 2020).

Another company focusing on distributing energy access across Africa is the Bithub, a blockchain accelerator for local start-ups, founded in 2015 and based in Kenya where the ICT sector is growing exponentially. The company aims at training 2000 blockchain engineers to develop Melanin Solar, a distributed solar energy solution, and other blockchain-based solution in Africa. The Bithub has begun engaging with the Kenyan government and regulators to advocate for the adoption of blockchain technology in Kenya's ICT policy (EY Global, 2019).

To sum up, the current situation on ICT and blockchain in Africa shows that the ingredients are in place to leapfrog the P2P energy trading using blockchain in SSA.

Policy and Regulatory Challenges

- **Restricted private participation in the mini-grid sector**

One of the major barriers to the development of a decentralized and distributed energy system in SSA is the restriction of the private sector to invest in electricity networks (figure 4). However, the private sector is better to operate in the mini-grid sector. The low private participation can be associated with the lack of comprehensive policies (Nkiriki & Ustun, 2017). According to the African Energy Outlook 2019, sixteen (16) out of 43 SSA countries do not allow the private sector to participate in electricity generation or sharing, and the other eighteen (18) countries allow only power generation (IEA, 2019). Transmission and distribution grids have monopolistic characteristics and are subject to severe regulation in many countries in the region whereas energy generation is more open. Nevertheless, in recent years, private investors showed interest in mini-grids and stand-alone systems due to progressively supportive policies and regulations and maturing markets as stated above.

- **The difficulty of the regulatory process and licensing requirements**

The regulatory process for the conventional mini-grid system is already expensive and time-consuming, so the regulatory process for the P2P energy sharing community would probably be more complicated since it is a new system. Requiring concessions, licenses and environmental approvals often delayed project developments.

Figure 2. Private sector participation in the electricity supply chain in SSA (IEA, 2019)

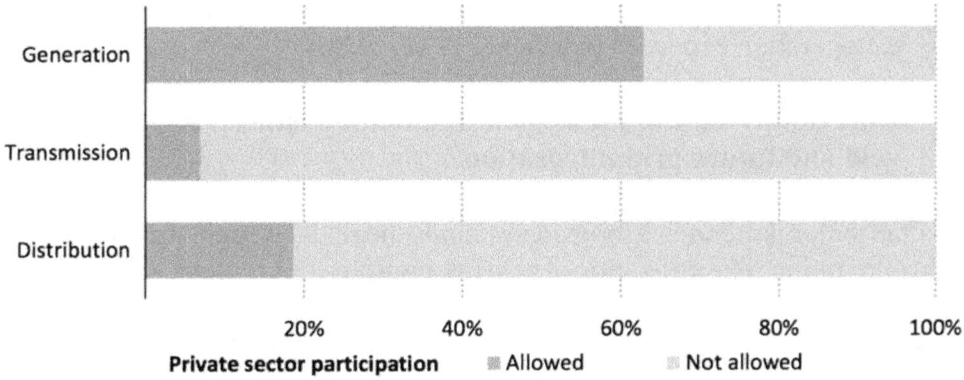

Some mini-grid projects are constrained to use temporary structures or have to limit their size and location to avoid the complicated formal requirements. There is also the possibility of combining projects with similar size and technical standards in one programmatic permit.

After reviewing the policy documents of several SSA countries, generally, licensing requirements depend on the installed capacity of the mini-grid. Although most small mini-grids are exempt from the license, requesting the exemption can be as difficult as requesting a full license, as is the case in Uganda. Nevertheless, some countries are more favorable, for instance in Rwanda, license or environmental clearance is not required for mini-grids under 500kW. In Tanzania, the processes are clearer and developers can proceed quickly and smoothly, licensing is not needed for less than 1MW and tariff approval is not required below 100kW. However, all projects must be registered with the EWURA (Energy and Water Utilities Regulatory Authority and get an environmental clearance) (EEP Africa, 2018).

Since the P2P energy community is a form of a mini-grid system, it should follow the same requirements.

- **Issues with the tariff framework**

According to the EEP portfolio projects in Africa, mini-grid developers are facing an issue with the tariff framework where regulatory bodies expect mini-grid developers to sell electricity at a similar price to the national grid (EEP Africa, 2018). The sale price doesn't take into consideration the additional costs that occurred with the bureaucracy and regulatory process.

The grid electricity price is often not cost-effective due to heavy subsidies from the governments. Therefore, for mini-grid developers to cover their costs, they have to secure grants or subsidies for their capital expenditure or also the operating expenses (Cabanero et al., 2020).

- **Uncertainty of the regulation about the connection to the main grid and future grid integration**

One major issue of a mini-grid system is the unclear regulation regarding grid integration. The utilization of grid integration would enhance the viability of P2P energy trading, as surplus energy generated within the community can be sold to the main grid. The question is how the energy market mechanism would function towards the existing main grid or the future grid connection. Some mini-grids are designed to be connected to the national grids since the inception of the project, in this case, they would negotiate PPAs (Power Purchase Agreement) with the grid utility to feed electricity into the main grid. However, the PPAs process can take a long time and some projects couldn't reach the implementation stage due to failure of finding a favorable agreement. If successful, this connection can provide financial sustainability to the mini-grid company (Nkiriki & Ustun, 2017).

The case of most P2P markets which would be developed in rural SSA is stand-alone grids. Because of this, severe challenges may arise due to the unpredictability of the grid extension plan. According to the EEP portfolio in Eastern and Southern Africa, several projects had to change location. Most of the rural electrification strategies in African countries do not provide details on how mini-grids will be connected to the national grid in the future when the main grid reaches the location. That situation often limits private investment within the sector. Countries like Tanzania have adopted new rules in 2017 which ensure compensation for private companies for the value of their mini-grid without the subsidies that they received. However, most other countries do not have specific guidelines or if they do, they have not been put into practice yet.

Opportunities for Promoting Digital Technology and Blockchain in SSA

- **Decreasing the cost of stand-alone solar technology**

Although the on-grid system has always been the main source of energy supply in Africa, the last recent years, decentralized solutions have gained a considerable increase in the region compared to the rest of the world. That is due to the decreasing cost of stand-alone solar PV and battery storage technologies combined with new business models using digital and appliance innovations which make the decentralized system more competitive. According to the African energy outlook 2019, mini-grid and stand-alone systems offer the least-cost solution to serve over 10% of electricity supply equalling around 160 TWh in Africa by 2040 (IEA, 2019).

Over the past few years, a boom of solar home systems with mobile payment and the PayGo mechanism has been observed in Sub-Saharan Africa. Innovative digital technologies and finance tools have a considerable impact on reducing the number of people without access to electricity in SSA. Currently, Africa has only 50 GW of renewable capacity whereas 36GW is from hydropower. Nevertheless, the recent years, many African countries have put effort into solar home systems, mini-grid, and micro-grid and the digitalization of communication and financial services has played a very critical role in the development of those new systems (IEA, 2019). Particularly, the development of telecommunication and payment infrastructures associated with the large increase in the availability of mobile phones and mobile money accounts is one of the most impactful forms of digitalization used in the energy sector (IEA, 2017)

Particularly for the Solar home system, the possibility of affordable payment plans over several months or years has contributed to the widespread of this technology in Africa. The customer pays only an initial deposit at the beginning, then a reasonable daily payment until the expiration of the payment. Mobile networks have allowed direct communication with customers and remote control of devices that disable the SHS when the customers were not able to pay. With such a mechanism, the ReadyPaysolar Systems of the Fenix International Company has enabled 500 000 households in SSA to benefit from their SHS. Like Fenix, the PowerCorner project is also a subsidiary of ENGIE which is the market leader in off-grid solar in Africa. The PowerCorner promotes smart mini-grids powered by solar energy with battery storage. It uses digital financial solutions like mobile money and Pay As You Go technologies to provide 24/7 energy services to households and local businesses in rural Tanzania and Zambia (Off Grid Energy Independence, 2019). Digital payment could facilitate as well the purchase of more efficient appliances. For instance, for solar TV, 272,485 units of TV were sold in SSA using the Cash and PAYGo mechanism (GOGLA, 2019).

The evolution of SHS is very considerable in SSA (figure 3), particularly in Eastern Africa. This will be favorable for the promotion of prosumer models.

Figure 3. Semi-annual evolution of large solar home systems sold regionally (GOGLA, 2019)

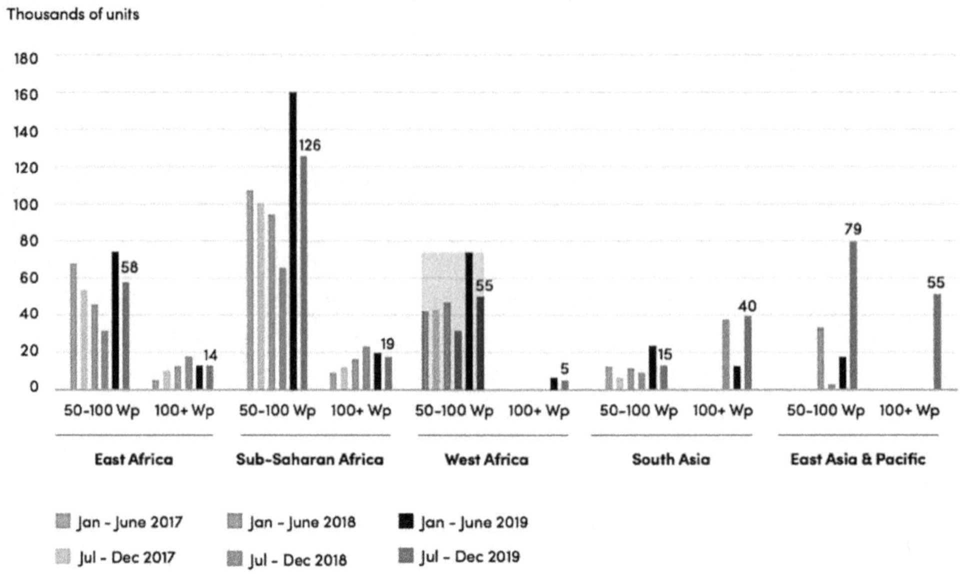

- **Expansion of Investment in the mini-grid sector in SSA**

According to the estimation of the World Bank, the amount of investment in mini-grids in SSA amounts so far to 4 billion USD for 1500 mini-grids (ESMAP, 2019). The majority of those mini-grids were financed and operated by state-owned utility companies. Nevertheless, private finance has been expanding in recent years and around 480 mini-grids in Africa today are developed by the private sector.

Regarding the global solar market and stand-alone systems, East and West Africa are particularly more advanced. Between 2012 and 2017, 75% of the funds raised by top developers account for developers operating in East and West Africa. It amounts to almost 700 million USD (World Bank, 2018).

- **The willingness of the African governments to improve policy and regulation in the mini-grid sector**

Recently, there is an increase in the understanding and awareness within SSA, particularly in Eastern and Southern Africa about the advantages of mini-grid in terms of reliability and socio-economic empowerment. Mainstreaming off-grid renewable energy in national rural electrification strategies was one of the key messages from the IOREC (International Off-grid Renewable Energy Conference & Exhibition) 2016 in Kenya. Therefore, creating an ecosystem that accelerates off-grid deployment is very essential, and dedicated policies and regulations is one of its foundation (IRENA, 2017).

Off-grid renewable energy solutions need to be introduced as early as possible in the region and it has to pass through the national electrification planning processes. Doing so would guide the public and private sector, as well as development banks and donors, to collaborate, mobilize, and direct resources to off-grid, and grid-based electrification options (IRENA, 2018). Some countries such as Nigeria, Rwanda, and Tanzania have recently incorporated mini-grid solutions in their energy plans and strategies to provide the basis for the expansion of electricity services.

International organizations are also taking part actively in promoting the mini-grid sector in SSA. For instance, the GEF or Global Environmental Facility has led a huge mini-grid program of 344,310,000USD called the GEF-7 Africa Minigrids Program. The expected outcome of the project is to promote innovation and technology transfer for sustainable energy breakthroughs for decentralized renewable power with energy storage. The first component consists of bringing technical assistance to Policy and Regulation to ensure that appropriate policies and regulations are in place including a mini-grid regulatory framework concerning tariff model, tax regime, and grid expansion risk (GEF, 2019).

- **The emergence of national and regional associations to lobby for favorable regulatory and policy frameworks for private sector mini-grid investment**

In recent years, many private mini-grid developers are joining national and regional associations to lobby for policy and regulatory frameworks that are favorable for the implementation of microgrid infrastructures. The associations play an important role in coordinating and closing the information gap between policymakers, practitioners, and investors. Examples of such associations are the Alliance for Rural Electrification (ARE), the African Mini-grids Developers Association (AMDA), and SEforALL Mini-Grids Partnership (MGP). They provide interesting platforms where private and public sectors can consult and collaborate on building enabling regulatory and financial frameworks (EEP Africa, 2018). Particularly, AMDA, or African Mini-grid Developers Association has 8 principles divided into two categories: regulatory and financial. The regulatory issues cover the permitting policies, the tariff framework, the grid integration framework, and the technical and safety standards. Emulating the work done by these associations for P2P energy trading and smart grids would help facilitate the promotion of this technology within Africa.

SOLUTIONS AND RECOMMENDATIONS

Required Policy

A favorable and adequate policy is necessary to promote P2P energy trading in SSA After analyzing the requirement of the P2P energy transaction system and the current situation of the policy in the region, four main areas of intervention have been identified (figure 5).

The first point is a supportive policy encouraging the decentralization of the power system. Therefore, it would be necessary that a well-elaborated mini-grid policy would be mainstreamed in the national energy policy, especially for the rural electrification plan. This will help to create an ecosystem that will encourage the public sector and attract the private sector to invest in the mini-grid while providing clear guidelines on engagement. It will encourage banks and donors as well to fund such projects.

The second point consists of encouraging a pilot program in P2P energy trading using blockchain. Since P2P energy systems and blockchain are both new concepts in SSA, engaging in investment in the sector could prove to be difficult as policy should be constantly reviewed investors may find it risky. That is why governments should support pilot programs as a test bed, in a regulatory sandbox that provides a framework to allow FinTech start-ups and other innovators in blockchain technology to conduct live experiments in a controlled environment under a regulator's supervision.

The third point which is an important requirement is to support access to capital for platform developers. As previously stated, the investment in the sector is new and it would be hard for start-ups to get funding for their projects, therefore governments can intervene to facilitate their access to financing. It can be through ensuring a guarantee, backing their project, or providing direct incentives by removing some taxes related to their activities.

One of the major concerns hindering the development of P2P energy trading and mini-grid solution, in general, is the difficulty of the regulatory process and license request. Therefore, bureaucracy related to that field should be limited and permit requests should be facilitated either in time or cost so that the project will be competitive.

Figure 4. Schematic overview of the policy framework

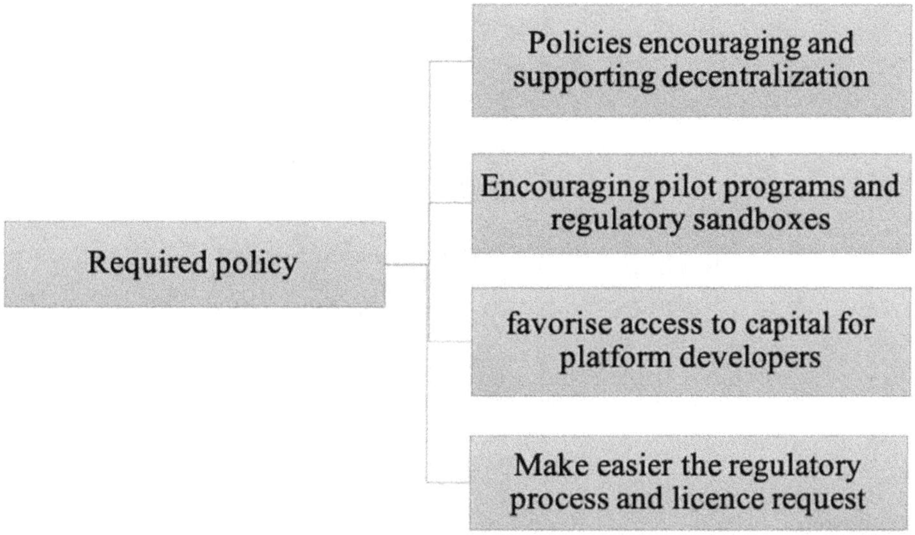

Regulatory Requirements

The regulatory requirement for P2P energy trading can be divided into two: the retail market and the distribution network (figure 6).

- Retail market regulation

The fundamental regulation required for P2P energy trading is to enable energy trading among prosumers and consumers within the community. It should consist in providing the right to sell electricity or precisely legislation permitting local P2P trading. It involves also defining well the statutes of prosumers and consumers and their respective roles and engagement.

Another requirement is to establish regulation on data collection and access as well as cybersecurity and privacy for the developers or platform owners and the peer participants. One feature of a blockchain is that all of the data and transactions on a blockchain are visible, this may threaten the privacy of the parties participating in the P2P community (Son et al., 2020). This presents the fundamental reason why having adequate regulation matters as P2P trading should be transparent, and accountable yet secure. Defining clear roles and responsibilities of stakeholders is another reason why regulation matters, which would be expounded on in a subsequent section.

Figure 5. Conceptual model for the P2P energy trading regulation

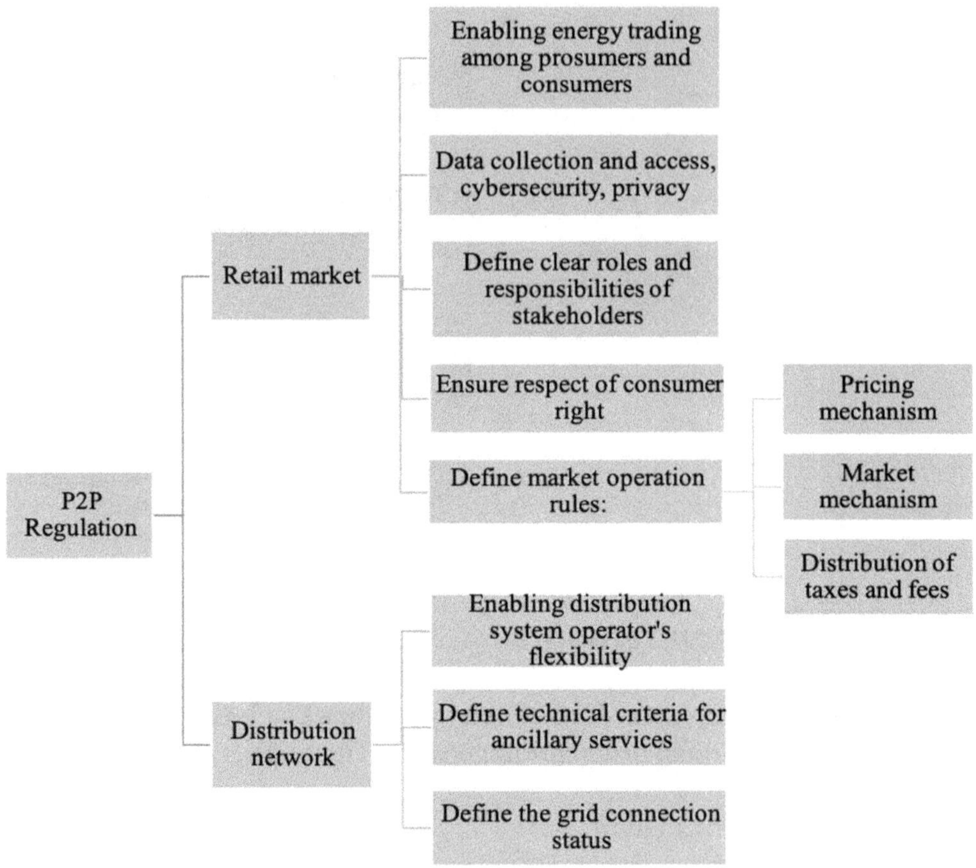

Furthermore, consumers are at the heart of P2P energy trading. Therefore, ensuring that their rights are respected among the stakeholders should be among the priorities. Consumers and prosumers need consumer protection. The aims of the regulation should be to provide a secure, affordable, consumer-friendly, efficient, and environmentally friendly supply of energy to the consumers. Some relevant legal principles have their basis in civil law but particularly, prosumers should be recognized in legislation and consumer law safeguard should be implemented in consumer-to-consumer transactions and their rights and obligations in P2P should be well-defined.

Then, market operation rules have to be very well-defined. It involves the pricing mechanism, the market mechanism as well as the taxation and fees.

- Distribution network

To procure flexibility from P2P platforms, enabling distribution system operators' flexibility is important.

It is necessary to define the technical criteria for ancillary services. The new ancillary services provided by a distributed energy system include inertial response, active power ramp, frequency response, voltage control, fault contribution, and harmonic mitigation (Oureilidis et al., 2020).

Defining the grid connection status is necessary to ensure the developers about what will happen when the main grid will reach the community where they operate. In a case where P2P trading is using the main grid, regulation is also needed to determine network charges.

Blockchain Framework in P2P Trading

Several elements, enabled by blockchain need to be regulated to have an efficient performance of the P2P energy system (figure 6).

Figure 6. Conceptual model for the P2P energy trading integrating blockchain

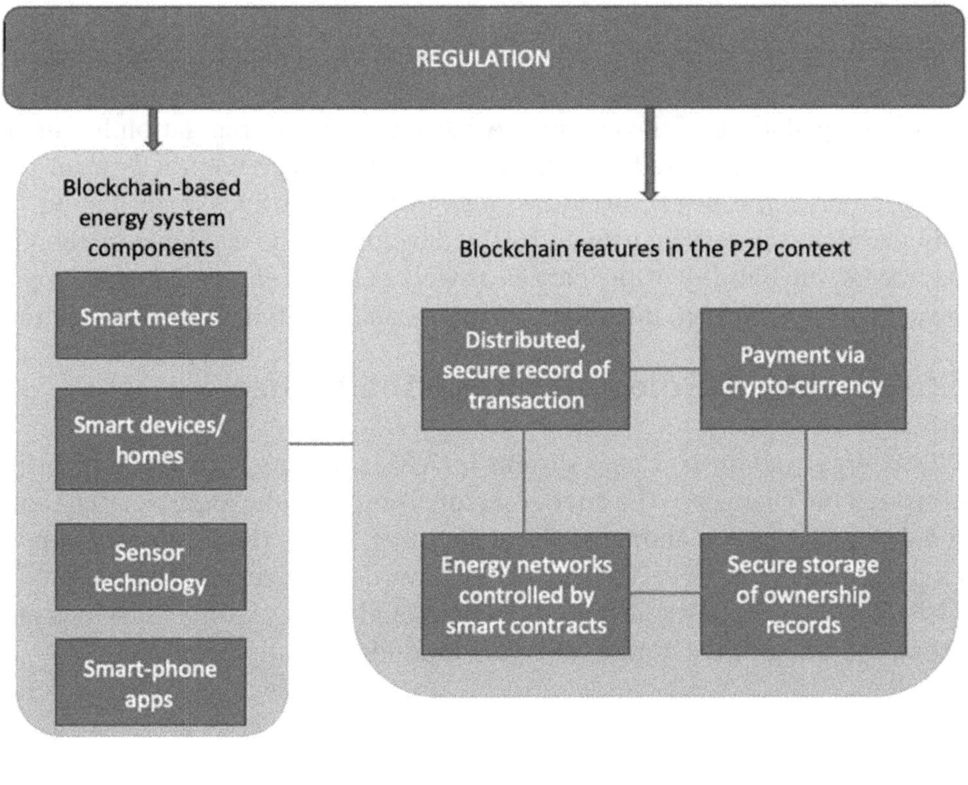

Blockchain technology is promising in enabling decentralized energy supply systems. It would enable the prosumers to sell the energy they generate directly to their neighbors. Blockchain systems initiate and transmit transactions while recording them in a tamper-proof manner. In a decentralized energy transaction and supply system, transactions between consumer and producer can be done automatically or manually and they are recorded in a tamper-proof way. The implemented policies and regulations should govern the different elements and processes within the system, from electricity generated via the network or the power grid to the consumer.

In a P2P energy system using blockchain, supply and demand are balanced by smart contracts including balancing market, microgrids, virtual power plants, and storage. To use blockchain systems, liberalizing the metering operation business is necessary, four main components are needed. They are smart meters, smart devices/homes, sensor technology, and smartphone apps. Smart meters, the crucial component provide intelligent measurement for measuring and transmitting the energy demand of consumers and the energy output of producers.

Although the concept of blockchain trading is new, the related legislation doesn't need to be ground-breaking. Some principles governing transactions using blockchain can be driven from the civil law principles or civil code of the country, but there can be any additional legal requirements for energy-related contracts. Innovative laws dealing with energy supply contracts need to be developed but be modeled after general civil law principles and other existing legislation, of the country. A basic framework regulation for smart contracts is also required (Oprea & Bâra, 2021). The contract's term has to be precise, the price variations, termination deadlines, information on any maintenance service offered by the developers, the available method of payments, the liability of the parties as well as any eventual damages to pay on a breach of contract, information system, and eventual dispute resolution.

Stakeholders' Roles and Responsibilities

P2P energy trading is a new system in Africa and its adoption will bring considerable changes to the energy sector. Therefore, the regulatory changes would involve a vast number of stakeholders across the electricity supply chain, with different levels of sensitivity towards the changes. The promotion of P2P energy trading will impact the national energy utility activities and those operating in the market of solar stand-alone equipment.

However, the main stakeholders involved directly in the P2P energy activities are the consumers & prosumers and the P2P market operators or developers in collaboration with ICT companies. The developed legislation needs to be clear to strike balance between consumer protection, the developers' benefits, and the community's impacts. For the prosumers and consumers, it is necessary to know how to manage people at the community level with respect to trading energy. Demarking distinct boundaries on prosumers' rights and obligations is also vital for implementing effective regulations as well as ensuring that local stakeholders are engaged.

It is necessary to recognize the crucial role played by local stakeholders, including public sector bodies in the context of P2P energy trading. African people have high regard for public figures, these figures can become a key intermediary between energy consumers and P2P market operators. Dialogues with industry and stakeholders involved in self-consumption projects are also recommended to learn from their experiences.

FUTURE RESEARCH DIRECTIONS

The present research has focused mainly on the policy and regulation framework study. However, the implementation of digital technology in the mini-grid sector, particularly the development of P2P energy trading involves many other aspects. Further study on those other aspects would be necessary to complete this research. It includes:

- Conducting assessments of the potential economic and social impacts of P2P energy trading using blockchain
- Conducting the technical study of the implementation of P2P energy trading and blockchain in an African context
- Building a business model and market design for blockchain-based P2P energy trading.

CONCLUSION

This research provided a diagnosis analysis of the policy and regulation opportunities and challenges for P2P energy trading using blockchain in SSA. There is a willingness of African governments and international organizations to improve policies and regulations in energy systems such as mini-grids and off-grid. Besides, the emergence of national and regional associations to lobby for favorable regulatory and policy frameworks for private sector investment in such energy systems is expected to improve the policy situation. However, there are still considerable restrictions on private participation within the energy sector. The regulatory process and licensing requirements can prove to be difficult with the issues with the tariff framework and the uncertainty of the regulation about the future grid integration.

The present research proposes a policy and regulation framework for the implementation of P2P energy trading using blockchain. To fully embrace all aspects involved in the system. General policy requirements, the regulatory requirement for P2P energy trading, and the blockchain framework in P2P energy trading were carried out separately in this study. The required basic policy should involve policies encouraging and supporting decentralization and also encouraging pilot programs and regulatory sandboxes. Access to capital for platform developers is also necessary and finally, it is important to make easier the regulatory process and license request for new developers who want to operate in the energy mini-grid sector including P2P energy trading. For the specific regulations of P2P energy trading, two blocs have to be considered, the retail market and the distribution network. The retail market involves enabling energy trading among prosumers and consumers as well as data collection and access, cybersecurity, and privacy. It is also necessary to define the clear roles and responsibilities of stakeholders and particularly ensure respect for consumer rights. The last which is very important is to define the market operation rules including pricing mechanism, market mechanism, and the distribution of taxes and fees. On the other side, the regulation of the distribution network consists in enabling the distribution system operator's flexibility, defining technical criteria for ancillary services, and defining the grid connection status. For integrating blockchain in the context of P2P energy trading, there should be clear regulation for all aspects where blockchain is involved such as the smart meters, smart devices, sensor technologies, and smart-phone applications. Clear regulation has also to manage and harmonize the blockchain features in the P2P context which are distributed and secure records of transactions, energy networks controlled by smart contracts, payment via cryptocurrency, and the secure storage of ownership records. Also, for the policy and regulations to be successful, the stakeholders' roles and responsibilities have to be very well defined. The present research can be used as a guideline for energy policy-makers of SSA countries that are still uncertain of the potential importance of embracing the disruptive technologies associated with the fourth industrial revolution mainly the use of blockchain. The study can also give a clear idea to investors and operators who are interested in operating in P2P energy trading and blockchain. Nevertheless, it is impossible to establish a "one size fits all" solution in this field. Although this research has provided the basic requirements for the policy and regulation framework for peer-to-peer P2P energy trading and blockchain, different aspects can be explored or some additional parameters may be required depending on the context of each country. Energy trading mechanisms and blockchain technologies are developing fast and have not slowed down, thus some adaptation or update may be required when using the proposed models in the future and the policy and regulation already established might be reconsidered to benefit from the technological and commercial changes. Establishing the policy and regulation in this sector involves a large number of interests and stakeholders which can present different sensitivity towards the change and it requires a harmonized concertation and agreement.

ACKNOWLEDGMENT

The authors are grateful to the African Union through the Pan African University for Water and Energy Sciences program, which provided funding and support for the research, and to George Mosomi from Rehub company in Kenya, who gave an insightful idea on the subject during his interview.

REFERENCES

AfDB. (2020). Developing Africa's Workforce for the Future. *Economic Outlook*, 2020.

EEP Africa. (2018). *Opportunities and challenges in the mini-grid sector in Africa: Lessons learned from the EEP portfolio*. EEP Africa.

Ahl, A., Yarime, M., Tanaka, K., & Sagawa, D. (2019). Review of blockchain-based distributed energy: Implications for institutional development. In *Renewable and Sustainable Energy Reviews* (Vol. 107, pp. 200–211). Elsevier Ltd. doi:10.1016/j.rser.2019.03.002

Alladi, T., Chamola, V., Rodrigues, J. J. P. C., & Kozlov, S. A. (2019). Blockchain in smart grids: A review on different use cases. In Sensors (Switzerland) (Vol. 19, pp. 4862). MDPI AG. doi:10.339019224862

Armbrust, M., Fox, A., Griffith, R., Joseph, A. D., Katz, R., Konwinski, A., Lee, G., Patterson, D., Rabkin, A., Stoica, I., & Zaharia, M. (2010). A view of cloud computing. In Communications of the ACM (Vol. 53, pp. 50–58). doi:10.1145/1721654.1721672

Bankymoon. (2020). *Blockchain enabled solutions and services*. http://bankymoon.co.za/

Blimpo, M. P., & Cosgrove-Davies, M. (2019). Electricity Access in Sub-Saharan Africa: Uptake, Reliability, and Complementary Factors for Economic Impact. World Bank & AFD.

Bobrow, D. B. (2006). Policy design: Ubiquitous, necessary and difficult. In *Handbook of Public Policy* (pp. 75–96). SAGE Publications Inc. doi:10.4135/9781848608054.n5

Cabanero, A., Nolting, L., & Praktiknjo, A. (2020). Mini-Grids for the Sustainable Electrification of Rural Areas in Sub-Saharan Africa: Assessing the Potential of KeyMaker Models. *Energies*, *13*(23), 6350. doi:10.3390/en13236350

Consensys. (2018). *Welcome to the Future of Energy- Grid+ White paper*.

Deing, J. (2019). *Peer-to-Peer Energy Trading Still Looks Like a Distant Prospect*. Greentech Media. https://www.greentechmedia.com/articles/read/peer-to-peer-energy-trading-still-looks-like-distant-prospect

Diestelmeier, L. (2019). Changing power: Shifting the role of electricity consumers with blockchain technology – Policy implications for EU electricity law. *Energy Policy*, *128*, 189–196. doi:10.1016/j.enpol.2018.12.065

Ecobank. (2018). *African crypto regulation*. Ecobank.

ESMAP. (2019). *Mini Grids for Half a Billion People: Market Outlook and Handbook for Decision Makers. Executive Summary*. ESMAP.

Ferguson, K. K., Soutter, L., & Neubert, M. (2019). Digital payments in Africa - how demand, technology, and regulation disrupt digital payment systems. *International Journal of Teaching and Case Studies*, *10*(4), 319. doi:10.1504/IJTCS.2019.103771

GEF. (2019). *GEF-7 Africa Minigrids Program*. GEF.

Global E. Y. (2019). *How blockchain can help to tackle sub-Saharan Africa's challenges*. (n.d.). EY. https://www.ey.com/en_sa/digital/tackling-sub-saharan-africa-s-challenges-with-blockchain

GOGLA. (2019). *Global Off-Grid Solar Market Report Semi-Annual Sales and Impact Data*. Gogla.

Gordon, P. (2019, July 16). Shell buys into LO3 Energy blockchain-based community platform. *Smart Energy News*. https://www.smart-energy.com/news/shell-buys-into-blockchain-based-community-energy-platform-lo3-energy/

IEA. (2017). Digitalization & Energy. In Digitalization & Energy. IEA. doi:10.1787/9789264286276-en

IEA. (2019). Africa Energy Outlook 2019. *World Energy Outlook Special Report*, 288.

IEA, IRENA, UNSD, World Bank, & WHO. (2020). *Tracking SDG 7: The Energy Progress Report 2020*. WHO.

Inês, C., Guilherme, P. L., Esther, M. G., Swantje, G., Stephen, H., & Lars, H. (2020). Regulatory challenges and opportunities for collective renewable energy prosumers in the EU. *Energy Policy*, *138*(April), 111212. doi:10.1016/j.enpol.2019.111212

IRENA. (2017). Accelerating Off-grid Renewable Energy. *Key Findings and Recommendations from IOREC*. IRENA.

IRENA. (2018). *Policies and regulations for renewable mini-grids*. International Renewable Energy Agency.

IRENA. (2019). *Innovation landscape brief: Blockchain*. International Renewable Energy Agency.

IRENA. (2020). *Peer-to-peer electricity trading: Innovation Landscape Brief*. International Renewable Energy Agency. www.irena.org

Jordan, E. A., Kusakana, K., & Bokopane, L. (2018). Prospective Architecture for Local Energy Generation and Distribution with Peer-to-Peer Electricity Sharing in a South African Context. *2018 Open Innovations Conference, OI 2018*, (pp. 161–164). IEEE. 10.1109/OI.2018.8535971

Khan, S. U. D. (2020). Environmental sustainability: a clean energy aspect versus poverty. *Environmental Science and Pollution Research, 28*(11), 13097–13104. doi:10.1007/s11356-020-11520-6

Khatoon, A., Verma, P., Southernwood, J., Massey, B., & Corcoran, P. (2019). Blockchain in energy efficiency: Potential applications and benefits. *Energies, 12*(17), 1–14. doi:10.3390/en12173317

Klein, L. P., Krivoglazova, A., Matos, L., Landeck, J., & de Azevedo, M. (2019). A novel peer-to-peer energy sharing business model for the Portuguese energy market. *Energies, 13*(1), 125. doi:10.3390/en13010125

Koffman, T. (2019). *Blockchain - Africa Rising*. Forbes. https://www.forbes.com/sites/tatianakoffman/2019/04/04/blockchain-africa-rising/#6e22abc7711f

Lightency. (2020). *Lightency-The Bright Side of Energy*. Lightency. http://lightency.io/#/

Ma, Z., Bloch-Hansen, K., Buck, J. W., Hansen, A. K., Henriksen, L. J., Thielsen, C. F., Santos, A. Q., & Jorgensen, B. N. (2018). Peer-to-Peer Trading Solution for Micrograds in Kenya. *IEEE PES/IAS PowerAfrica*, (pp. 420–425). IEEE. doi:10.1109/PowerAfrica.2018.8520980

Majeed Butt, O., Zulqarnain, M., & Majeed Butt, T. (2021). Recent advancement in smart grid technology: Future prospects in the electrical power network. *Ain Shams Engineering Journal, 12*(1), 687–695. doi:10.1016/j.asej.2020.05.004

McKenzie, B. (2018). *Blockchain and Cryptocurrency in Africa.*

Mengelkamp, E., Gärttner, J., Rock, K., Kessler, S., Orsini, L., & Weinhardt, C. (2017). Designing microgrid energy markets: A case study: The Brooklyn Microgrid. *Applied Energy, 210*, 870–880. doi:10.1016/j.apenergy.2017.06.054

Morstyn, T., Farrell, N., Darby, S. J., & McCulloch, M. D. (2018). Using peer-to-peer energy-trading platforms to incentivize prosumers to form federated power plants. In Nature Energy (Vol. 3, pp. 94–101). Nature Publishing Group. doi:10.103841560-017-0075-y

Munsing, E., Mather, J., & Moura, S. (2017). Blockchains for decentralized optimization of energy resources in microgrid networks. *IEEE Conference on Control Technology and Applications (CCTA), 2017-Janua*, (pp. 2164–2171). IEEE. 10.1109/CCTA.2017.8062773

Mylrea, M., Nikhil, S., & Gourisetti, G. (2017). *Blockchain for Smart Grid Resilience: Exchanging Distributed Energy at Speed.* Scale and Security.

Ndung'u, N., & Signé, L. (2020). The Fourth Industrial Revolution and digitization will transform Africa into a global powerhouse. In Capturing the Fourth Industrial Revolution: A regional and national agenda (pp. 61–73). Foresight Africa 2020 report.

Nieman, A. (2015). *A FEW SOUTH AFRICAN CENTS ' WORTH ON BITCOIN. 18*(5).

Nkiriki, J., & Ustun, T. S. (2017). Mini-grid policy directions for decentralized smart energy models in Sub-Saharan Africa. *2017 IEEE PES Innovative Smart Grid Technologies Conference Europe, ISGT-Europe 2017 - Proceedings, 2018-Janua*, (pp. 1–6). 10.1109/ISGTEurope.2017.8260217

Noor, S., Yang, W., Guo, M., van Dam, K. H., & Wang, X. (2018). Energy Demand Side Management within micro-grid networks enhanced by blockchain. *Applied Energy*, *228*, 1385–1398. doi:10.1016/j.apenergy.2018.07.012

OECD. (2019). The Policy Environment for Blockchain Innovation and Adoption: 2019 OECD Global Blockchain Policy Forum Summary Report. *OECD Blockchain Policy Series*. OECD. www.oecd.org/finance/2019-OECD-Global-Blockchain-Policy-Forum-Summary-Report.pdf

Off Grid Energy Independence. (2019). *ENGIE acquires Mobisol, becomes market leader in off-grid solar Africa.* Off Grid Energy Independence. https://www.offgridenergyindependence.com/articles/18086/engie-acquires-mobisol-becomes-market-leader-in-off-grid-solar-africa

OneWattSolar. (2020). *Blockchain Powered Digital Electricity - OneWattSolar*. OneWattSolar. https://onewattsolar.com/

Oprea, S. V., & Bâra, A. (2021). Devising a trading mechanism with a joint price adjustment for local electricity markets using blockchain. Insights for policy makers. *Energy Policy*, *152*(February), 112237. doi:10.1016/j.enpol.2021.112237

Oureilidis, K., Malamaki, K. N., Gallos, K., Tsitsimelis, A., Dikaiakos, C., Gkavanoudis, S., Cvetkovic, M., Mauricio, J. M., Ortega, J. M. M., Ramos, J. L. M., Papaioannou, G., & Demoulias, C. (2020). Ancillary services market design in distribution networks: Review and identification of barriers. *Energies*, *13*(4), 917. doi:10.3390/en13040917

Paribas, B. N. P. (2019). *Blockchain is more than just numbers for these small farmers*. CIB. https://cib.bnpparibas.com/sustain/blockchain-is-more-than-just-numbers-for-these-small-farmers_a-3-3149.html

PLAAS. (2018). *Supply Chain Research Architecture*. PLAAS.

PricewaterhouseCoopers. (2016). Blockchain – an opportunity for energy producers and consumers? *Pwc.Com*, 1–45.

Rao, P. (2018). Africa could be the next frontier for cryptocurrency. *Africa Renewal*, *32*(1), 27–27. doi:10.18356/f6b3e553-en

Samuel, O., Almogren, A., Javaid, A., Zuair, M., Ullah, I., & Javaid, N. (2020). Leveraging blockchain technology for secure energy trading and least-cost evaluation of decentralized contributions to electrification in sub-Saharan Africa. *Entropy (Basel, Switzerland)*, *22*(2), 226. doi:10.3390/e22020226 PMID:33286000

Solshare. (2020). *ME Solshare*. https://me-solshare.com/

Son, Y.-B., Im, J.-H., Kwon, H.-Y., Jeon, S.-Y., & Lee, M.-K. (2020). Privacy-Preserving Peer-to-Peer Energy Trading in Blockchain-Enabled Smart Grids Using Functional Encryption. *Energies*, *13*(6), 1321. doi:10.3390/en13061321

Tushar, W., Saha, T. K., Yuen, C., Smith, D., & Poor, H. V. (2020). Peer-to-Peer Trading in Electricity Networks: An Overview. *IEEE Transactions on Smart Grid*, *11*(4), 3185–3200. doi:10.1109/TSG.2020.2969657

Tushar, W., Yuen, C., Mohsenian-Rad, H., Saha, T., Poor, H. V., & Wood, K. L. (2018). Transforming energy networks via peer-to-peer energy trading: The potential of game-theoretic approaches. *IEEE Signal Processing Magazine*, *35*(4), 90–111. doi:10.1109/MSP.2018.2818327

van Soest, H. (2019). *Peer-to-peer electricity trading: A review of the legal context*. Competition and Regulation in Network Industries., doi:10.1177/1783591719834902

Wongthongtham, P., Marrable, D., Abu-Salih, B., Liu, X., & Morrison, G. (2021). Blockchain-enabled Peer-to-Peer energy trading. *Computers and Electrical Engineering, 94*(September 2020), 107299. doi:10.1016/j.compeleceng.2021.107299

World Bank. (2018). Off-grid Solar Market Trends Report 2018. World Bank. doi:10.1017/CBO9781107415324.004

Zhang, C., Wu, J., Zhou, Y., Cheng, M., & Long, C. (2018). Peer-to-Peer energy trading in a Microgrid. *Applied Energy, 220*(June), 1–12. doi:10.1016/j.apenergy.2018.03.010

Zhou, Y., Wu, J., Long, C., & Ming, W. (2020). State-of-the-Art Analysis and Perspectives for Peer-to-Peer Energy Trading. *Engineering (Beijing), 6*(7), 739–753. doi:10.1016/j.eng.2020.06.002

KEY TERMS AND DEFINITIONS

Blockchain: A distributed and immutable ledger that stores data in blocks and are linked together via cryptography.

Decentralized Energy: Energy that is generated off the main grid.

Digital Technologies: All electronic tools and devices and automatic systems that generate store or process information or data. It includes the Internet of Things, Artificial intelligence, and blockchain.

Distributed Energy: Production of electricity by a large number of small generators which can be solar roofs or small wind turbines.

Mini-Grid: an electricity distribution network involving a small-scale electricity generation having a power rating of less than 15MW and disconnected from national electricity transmission networks.

Peer-To-Peer Energy Trading: A business model, based on an interconnected platform, that serves as an online marketplace where consumers and producers meet to trade electricity directly, without the need for an intermediary.

Policy: A set of ideas or plan for action adopted and implemented by any government entities to govern and rule one sector or particular situation.

Regulation: A set of rules, laws, directives, requirements, and restrictions adopted and applied by a government authority in order to manage and control a specific sector, practice, or situation.

Chapter 7
Tracking Public Financing of Adaptation Projects for Developing Economies Using a Climate Budget Tagging Framework for Nigeria

Chukwuemeka Onyebuchi Onyimadu
National Institute of Legislative and Democratic Studies, Nigeria

Daniel Uche Sunday
Michael Okpara University of Agriculture, Umudike, Nigeria

ABSTRACT

There is ample evidence in the literature that developing countries would suffer the most from the adverse effects of climate change. Although, respective developing economies have dedicated action plans to mitigate or adapt to these adverse effects, financing for these strategies may be lacking or national governments may not commit financial resources to actualizing these strategies. Using a budget analysis and climate budget tagging framework, the chapter evaluates the financial resources the Nigerian government has committed to its adaptation strategies as stipulated in the 2011 National Adaptation Strategy and Plan of Action on Climate Change (NASPA – CCN). The study found out amongst others that government expenditure on climate change tends to be more of mitigation than adaptation. In addition, adaptation programs targeted at the industry, commerce, telecommunications, and transport sector are most neglected among other sectors highlighted as priority sectors in the NASPA – CCN policy.

DOI: 10.4018/978-1-7998-8638-9.ch007

INTRODUCTION

For developing economies, adaptation to climate change challenges is now recognized as a key factor in determining the future outcomes of climate change impacts and mitigation strategies (Lobell, 2008). The literature on the challenges of climate change reveals that developing countries are more susceptible to risks arising from climate change. The reasons put forward are that these countries do not have the adaptive capacity to limit these risks (Moser & Ekstrom, 2010), and have economies that tend to depend greatly on climate change sensitive sectors (Lim, et al., 2004). Thus, a majority of the literature for developing economies focus on providing evidence based research on the socio – economic and political factors that limit adaptation strategies while controlling for the idiosyncrasies of each developing country.

In defining climate change adaptation, one must recognize that adaptation, as a concept is not strictly exclusive to issues bordering on climate change. A conventional working definition by the International Panel on Climate Change (IPCC) defines adaptation as, "the adjustment in natural or human systems in response to actual or expected climatic stimuli or their effects, which moderates harm or exploits beneficial opportunities" (Lobell, 2014). Moser and Ekstrom (2010) defined adaptation to include changes in the social – ecological system to anticipated and actual impacts of climate change, with these changes ranging from short term coping mechanisms to long term structural transformations. From both definitions, climate change adaptation implicitly implies substantive changes in consumption and production processes, which may have significant welfare implications and require deliberate involvement in planning for expected outcomes, scope and scale of adaptation strategies.

Therefore, in managing the accompanying risks of climate change, policymakers, prior to the first decade of the 21st century, focused mainly on strategies aimed at mitigation rather than adaptation (New et al., 2011). The emphasis then, was to reduce the potential negative size of climate change effects on the social – ecological system (Stafford et al., 2011). However, climate change challenges are time bounded. While the effects of climate change occur in the future – long term, the time bounds of mitigation strategies are often linked with the office term limits of decision makers and public servants – mostly four years and in the short term. Consequently, this time lag ensures that current mitigation strategies may not prevent the occurrence of future climate change challenges (Adger 2006). This necessitates the focus of climate change adaptation strategies.

There is adequate evidence on the negative impacts of climate change in Nigeria. NIMET (2008) noted significant weather related disasters with the expectation of a continued trend. Also observed is the high predisposition of Nigeria's natural and agricultural systems to climate change, with recorded incidence of flood and draught in the past four decades. To effectively respond to these challenges, the Nigerian government developed the National Adaptation Strategy and Plan of Action on Climate Change (NASPA – CCN) in 2011. This plan is designed as an integrated component of sustainable development, with the goal of reducing climate change vulnerabilities and impacts; improving adaptive capacities; leveraging new opportunities and facilitating stakeholder's collaboration. Key in achieving these objectives is the funding responsibility of the Federal Government of Nigeria, as mandated in the NASPA – CCN. Effective implementation of the policy overtime is expected to significantly tackle the already existing challenges of climate change, while adequately preparing the social – ecology system for future challenges.

Assuming the federal government considers as a priority, the objectives of the NASPA – CCN, the federal governments' (public) budgets should, in turn, reflect the prioritization of these objectives. This assertion becomes the basis for a critical assessment of government's ability to fulfil its funding responsibilities of the adaptation policy. Consequently, the paper intends to match adaptation strategies to the budgetary commitments by tracking and characterizing climate – change adaptation expenditures in Nigeria's budgetary system. Using the Appropriation Acts of 2013 through 2020, the chapter provides answers to the following questions; (1) what type of adaptation strategies has the federal government been committed to implementing? (2) Evaluate the nature of adaptation interventions in the budget in comparison with those stipulated in the NASPA – CCN policy? and (3) Is expenditure on climate change adaptation interventions reflect an effort towards progressive achievements? Progressive achievement here reflects the assumption that the budget is a developmental tool and should reflect quantifiable developments in the adaptation strategies over time.

BACKGROUND

The associated challenges of climate change is well documented in the literature (World Bank, 2010). What is of most concern to researchers and policy makers is, how these challenges affect countries unevenly, and the adaptation and mitigations strategies that can alleviate these challenges. According to UNDP

(2007), the emphasis on mitigation and adaptation strategies is paramount, as it is the poorest countries – especially countries in Sub – Saharan Africa – that are most vulnerable to these climate change challenges. Thus, to successfully adapt, decision makers in these poor economies are faced with limited options and constraints, which could reduce the potential damage of climate change.

From the literature, some of the constraints faced by policy makers include; inadequate information, characterized by a dearth of easily available data (Ford et al, 2016; Ziervogel, Johnson, Matthew, & Mukheibir, 2010); the complexity of planning for adaptation, considering seasonal and annual variabilities that require a wide range of policy response (Mukheibir, Kuruppu, Gero, & Herriman, 2013); existence of policy gaps due to dominance by central governments, weak institutions and non-inclusiveness of adaptation strategies (Ampaire et al., 2017); the difficulty in monitoring progress in adaptation due to conflicting metrics for measurement (Brooks et al., 2013; Ford et al., 2015); the time differences between success of adaptation and adaption program timescales(Adger 2006; Challinor et al., 2007); and needed clarity between reduced vulnerability, enhanced adaptive capacity and improved resilience, which are different targets for adaptation strategies (Thorarinsdottir & de Bruin, 2016).

A major source of challenge to adaptation strategies for developing economies is funding. Although the returns of a successful adaptation is substantial, limited financial resources for developing economies may constrain the effective implementation of current and future adaptation measures (UNEP, 2016; UNDP, 2016b). The World Bank expects a financial cost of $10 – 40 billion annually for developing economies and three times that amount by the UN Framework Convention on Climate Change (UNFCCC) (UNFP, 2009), with Fankhauser (2010) arguing that these cost outlays are grossly underestimated. Recent data from UNEP (2016) estimates $1140 – 300 billion annually, with the possibility of reaching $500 billion by 2050. These figures are only indicative of the financial needs of developing countries, with existing conceptual and practical challenges in actually tracking adaptation financing, especially in developing economies. These challenges are due to a lack of consistency and quality data as well as the complexity in differentiating between mitigation and adaptation financing (UNFCCC, 2018; CPI, 2018).

Although the flow of financial resources towards adaptation, mostly stems from developed to developing economies (UNFPA, 2009), sources of adaptation finance could vary between individual country's commitment to financing (through the budget), to international sources of financing and private or public institutions. International financial sources like the UNFCCC's dedicated climate change fund Global Environment Facility (GEF), have been criticized based on its inadequacies in providing adequate funding; governance structure that undermines ownership of the funds by low income countries; and a high transaction costs associated with these funds (Ayers, 2009). Research has also shown that, although there are a wide ranging multilateral, bilateral, and DFIs funding available (ODI, 2017), developing countries face significant obstacles in planning for, assessing and delivering climate finance, even when dealing with multilateral climate funds (Burmeister, Cochu, Hausotter, & Stahr, 2019). Also, the various requirements for accessing these international source of funds can be demanding, often requiring stringent conditions. This has led to the consideration of domestic adaptation finance sources.

The UNDP (2015) listed some of these domestic sources to include finance from national and sub – national governments. These finances could be raised through specific carbon taxes or provided for in budgets. From the literature on climate change adaptation finance, there is a limited focus on financing from national governments, with a preference in the literature for international financing (CPI, 2017). Despite the literature leaning towards international financing, the Climate Public Expenditures and Institutional Review (CPEIR) has provided a tool for tracking domestic expenditure on climate change adaptation. This tool allows for the use of both qualitative and quantitative methods in analyzing the structure, trend and priorities given to adaptation strategies. Tracking domestic adaptation expenditures is an important process of prioritizing allocation and identifying financing gaps. It provides the benchmark for assessing expenditures as well as advocating for further expenditures (CPI, 2018).

Resch et al., (2017) and Buckley (2014) both agreed on the need for developing the capacity of developing economies to track adaptation finance. This ensures the quality of data and the possible integrating of adaptation strategies in respective country's national planning documents. Given that adaptation finance in developing economies is likely to be from domestic public finance, there is a need for a focus on adaptive methodologies for tracking these finances. Resch et al., (2016) provided some basic methodologies which developing economies may use for tracking adaptation finance; Budget analysis, Public Expenditure Review, and Budget tagging. As adapted from Resch et al. (2017), Table 1 provides a comparison of these different adaptation expenditure-tracking methodologies.

Table 1. Comparisons of various methodologies for tracking adaptation expenditure

Features	Budget Analysis	Public Expenditure Review	Budget Tagging
Easy to Implement (does not require capacity building)	Positive	Negative	Positive
Quick	Positive	Negative	Positive
Cost Effective	Positive	Negative	Neutral
Standardization (allows for cross - country comparison)	Neutral	Positive	Neutral
Encompasses full budget cycle	Negative	Negative	Positive
Can be integrated into budgeting process	Positive	Negative	Positive
Independent from timely publication of budget data	Negative	Negative	Positive
independent from sufficiently disaggregated data	Negative	Negative	Neutral
Enables comparison of different levels & composition of expenditures against objectives	Negative	Positive	Negative

Source: Adapted from Resch et al., (2017). Note, positive, neutral and negative describes the advantages (positive) and disadvantages (negative) each methodology given a feature category.

Resch et al. (2017) defined budgetary analysis as an approach that covers budgetary proposal and actual expenditures. Developing countries often use this methodology due to its low cost. However, the available studies that have employed this methodology focused on South Asia (ACT, 2018), Afghanistan and India: Kerala, Chhattisgarh, and Bihar (Allan et al. 2016). In using the public expenditure methodology, Micale, Tonkonogy, & Mazza (2018) criticized it based on being time consuming and expensive. Nevertheless, they noted that the method is normally used at the national and sub – national government level. The method has been applied in Bangladesh, Cambodia, China, Fiji, Kiribati, Morocco, Indonesia, Nepal, Thailand, Tonga, Vietnam and Pakistan (Resch et al. 2017; UNDP 2016a, 2015 and 2012). The budget tagging methodology is more precise in capturing allocation weaknesses in tracking adaptation financing. It is an approach that flags budget codes that are relevant to adaptation in the government's financial system (Micale, Tonkonogy, & Mazza, 2018). This method has been used for the Philippines, Nepal, Indonesia and Bangladesh (Micale, Tonkonogy, & Mazza, 2018; UNDP, 2015).

With respect to broader issues that affect financing adaptation financing, Barr et al (2010) backed the allocation of adaptation finance in a transparent, efficient and equitable way in order to ensure best returns from financing adaptation strategies. They argued that adaptation finance should be allocated based on the climate change impacts experienced in a country, a country's adaptive capacity and its implementation capacity. Some scholars have also opined that the effects of adaptation strategies on the adverse impacts of climate change, is similar when countries promote pro – poor inclusive growth (Bowen, Cochrane, & Fankhauser, 2012). This implies that a lower adaptive capacity present in developing economies would in most cases; limit the expected positive impact of adaptation strategies. Thus, Barr et al. (2010) argued that adaptation strategies would be more effective in such counties when strategies that improve institutions, health sector, education and the financial sector – where all these sectors are closely linked to inclusive growth. Thus, financing for adaptation strategies, would implicitly involve financing pro – poor growth (Dodman, Ayers and Huq, 2009).

For developing economies, the agricultural sector plays a dominant role in reducing the incidence of poverty. However, the agricultural sector is most vulnerable to climate change. Barry and Skinner (2002) noted that adaptation strategies in the agriculture sector should include the financing of technological developments, government programs and insurance, innovative farm production practices and farm financial management. Financing these aspects would ensure that adaptation strategies are effective. However, there is evidence that financing these adaptation strategies are constrained by the option between potential adaptation options, government decisions that direct financial resources to adaptation strategies and risks associated to loss of limited financial resources.

In summary, the reviewed literature showed that developing countries are increasingly turning away from international sources of financing climate – change adaptation strategies to domestic public financing. A major reason for this is the relative difficulty associated with these international financing sources. However, a common conclusion from the literature is the importance of having adequate financial resources, and the need for tracking, monitoring and tagging all expenditure on adaptation. This places emphasis on the possibility of identifying financial gaps and prioritizing expenditure for adaptation. On the three methodologies commonly used to assess financing for adaptation, there are limited studies in the literature that integrated these methods, with most studies using any of each method on a standalone basis. In addition, the use of these methodologies in developing economies lean more towards countries in South Asia and Middle East. The study was unable to find any documented use of these methodologies for Nigeria.

A CLIMATE BUDGET TAGGING FRAMEWORK (CBT)

The paper employs a hybrid methodology, which applies OECD Development Assistance Committee (DAC) (OECD, 2011) Climate Budget Tagging (CBT) tool in a Budget Analysis Framework[1]. In employing this methodology, the study links climate – change adaptation strategies to Federal Government's budgetary commitments to its climate change adaptation policy. The CBT allows for the compilation of comprehensive data on climate change spending, enabling policy makers to make informed decisions and prioritize climate investments (Le, 2015). The CBT is also provides some advantage in its provision for public scrutiny of government's spending towards addressing climate change issues. Following the guidelines in OECD-DAC (2011) and EBRD (2014), for the effective use of the CBT tool in a budget analysis framework, the following components of the CBT must first be established; (a) Definition of climate change adaptation activities/markers; (b) Classification of climate change adaptation expenditures; (c) Weighing of relevance of adaptation interventions; and (d) Designing the tagging procedure. OECD – DAC (2011) provides clear definitions on the definitions and classifications (see Table 2) and the scoring/weighing and tagging procedures (See Figure 1 and Table 3). Detailed analysis using data from the four components of the CBT will provide answers to questions 1 and 2.

Table 2. Climate change adaptation markers and classification of expenditures

Definition of activities; An activity should be classified as adaptation related, if:	**Clarification;** It intends to reduce the vulnerability of human or natural systems to the impacts of climate change and climate-related risks, by maintaining or increasing adaptive capacity and resilience. This encompasses a range of activities from information and knowledge generation, to capacity development, planning and the implementation of climate change adaptation actions.
Eligibility of Activity; An activity is eligible for the climate change adaptation marker if;	**Clarification:** a) the climate change adaptation objective is explicitly indicated in the activity documentation; and b) the activity contains specific measures targeting the definition above. Carrying out a climate change adaptation analysis, either separately or as an integral part of agencies' standard procedures, facilitates this approach.

Source: OECD – DAC (2011)

The adaptation markers and scoring format aims at monitoring financial flows to adaptation programs, whose objectives are specifically aimed at climate change adaptation. The weighing/scoring schema provides in figure 1, uses questions to track financing to policy objective. A principal objective is weighted higher than other objectives, when the funded adaptation activity would not have been funded, if not for its objective. Conversely, a significant objective is scored lower than a principal objective, when an adaptation activity has other objectives, but these objectives are formulated to help meet adaptation concerns. Financing figures arrived at by the marker/activity classification and the scoring system for both principal and significant objectives are regarded as estimates (an upper bound) of adaptation financing (OECD – DAC, 2011).

To derive the scores for adaptation programs, the study created a scoring procedure that acts as a guide when scrutinizing each Ministry, Agency, and Department (MDAs). The scoring procedure (Table 3) matches each MDAs budgetary allocations to targeted sectors that have identified by the National Adaptation Strategy and Plan of Action on Climate Change as beneficiaries of adaptation programs. For example, for the Ministry of Health, it is intuitive to expect that adaptation activities captured in the Ministry's budget would have both principal and significant objectives in agriculture, fisheries, biodiversity, freshwater resources etc. We limit our scrutiny of budgets to the periods 2013 – 2020. The choice of starting from 2013 is justified on the basis that the National Adaptation Strategy and Plan of Action on Climate Change was introduced in 2011. Although the budgets of 2015 and 2016 were included in the period of interest, the study recognizes that Nigeria was in a recession in 2015 and the budget for 2016 is expected to prioritizes economic recovery programs over climate change adaptation programs.

The second half of our hybrid methodology involves the use of a budget analysis framework. The chapter employed this framework to determine if there has been progressive achievements in the federal government's commitment to financing adaptation strategies. The total budgetary expenditure on adaptation strategies was collated, adjusted for inflation and disaggregated by priority sectors (agriculture, water resources, forests, biodiversity, health and sanitation, human settlements and housing, energy, transport and communication, industry and commerce, disaster, migration and security, livelihood, vulnerable groups, and education) as stipulated in the NASPA – CCN policy framework. This approach will provide answers to question 3.

Figure 1. Weighing/scoring system for adaptation markers/activity
Source: OECD – DAC (2011)

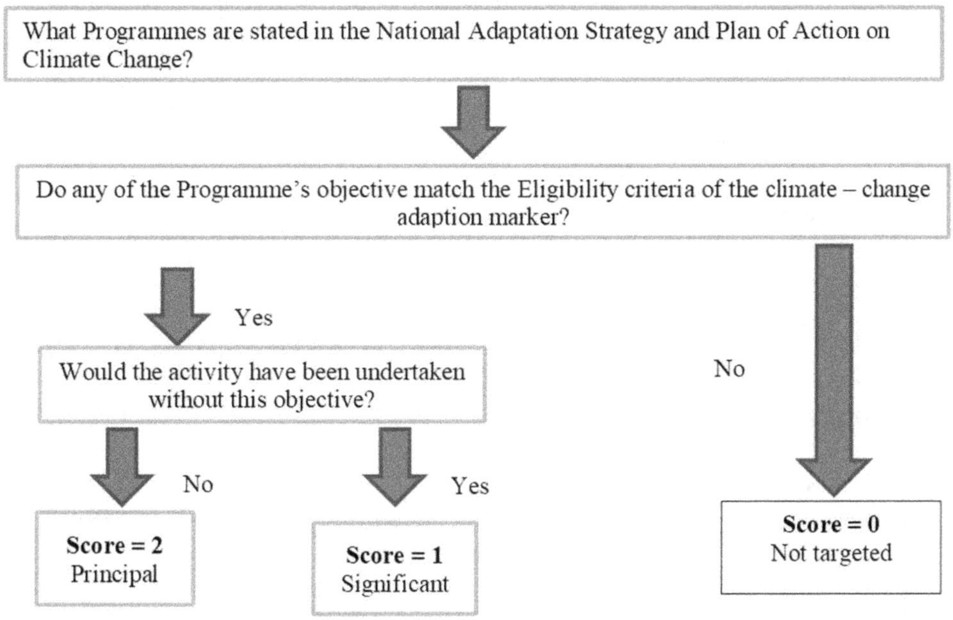

The use of such budget analysis framework is not without its limitations. Political and philosophical questions cannot be answered using this methodology. Following Schuftan (2005) arguments, a budget analysis framework may provide answers to question of what has been or what is being spent by the government, but cannot answer questions of what should be spent. Additionally, effectiveness and efficiency in the implementation of adaptation activities cannot be determined using this methodology. Proponents of this method strongly advocate for the use of complementary sectoral information and actual field observations on the operationalization of the budget when drawing inference (Schuftan, 2005). Another important limitation on the application of this method to climate change adaptation financing, is the dearth of quality data on actual expenditures. The authors circumvent this problem by using budgetary allocations to adaptation activities as a measure of financial commitment.

Table 3. Expected scoring procedure for MDAs

MDAs	Budget ID	Project Title	Status	Amount	Agriculture (Crops and Livestock)	Freshwater Resources, Coastal Water Resources and Fisheries	Forests	Biodiversity	Health and Sanitation	Human Settlements and Housing	Energy	Transport and Communications	Industry and Commerce	Disaster, Migration and Security	Livelihood	Vulnerable Groups	Education
Agriculture	X	X	New/Ongoing		X	X	X	X	X	X			X	X	X	X	
Aviation	X	X	New/Ongoing									X	X		X		
Education	X	X	New/Ongoing		X	X	X	X	X	X				X	X	X	X
Energy	X	X	New/Ongoing		X	X	X	X		X	X	X	X		X	X	
Environment	X	X	New/Ongoing		X	X	X	X	X	X	X	X	X	X	X	X	X
Health	X	X	New/Ongoing		X	X	X	X	X	X				X	X	X	X
Information and Culture	X	X	New/Ongoing		X	X	X	X	X	X	X			X		X	X
Interior	X	X	New/Ongoing						X			X	X	X	X	X	
Labour and Productivity	X	X	New/Ongoing		X					X	X	X	X			X	
Mines and Steel Development	X	X	New/Ongoing				X	X			X	X	X	X	X		
Tourism, Culture and National Orientation	X	X	New/Ongoing							X		X	X	X	X		
Niger Delta	X	X	New/Ongoing		X	X	X	X	X	X	X		X	X	X		

MDAs	Budget ID	Project Title	Status	Amount	Agriculture (Crops and Livestock)	Freshwater Resources, Coastal Water Resources and Fisheries	Forests	Biodiversity	Health and Sanitation	Human Settlements and Housing	Energy	Transport and Communications	Industry and Commerce	Disaster, Migration and Security	Livelihood	Vulnerable Groups	Education
NDDC	X	X	New/Ongoing		X	X	X	X	X	X	X		X	X	X		
Ecological Fund	X	X	New/Ongoing		X	X	X	X	X	X	X		X	X	X		
Petroleum Resources	X	X	New/Ongoing		X	X	X	X	X	X	X	X	X	X	X	X	
Works and Housing	X	X	New/Ongoing						X	X	X	X			X	X	
Power	X	X	New/Ongoing						X	X	X	X	X		X	X	
Water Resources	X	X	New/Ongoing		X	X	X	X	X	X				X	X	X	X
Women Affairs	X	X	New/Ongoing		X	X	X	X	X	X				X	X	X	X

Source: Adapted from OECD – DAC (2011)

APPLICATION OF CBT TO CLIMATE CHANGE ADAPTATION FINANCING IN NIGERIA

A starting point for providing answers to the questions posed in this paper is the collation of climate change financial data from budget documents of various years, and differentiating between mitigation and adaptation programs in these budgets. For the later, the paper depends on the OECD (2011) CBT's definition of what programs constitute mitigation and adaptation. According to OECD (2011), programs are tagged mitigation if programs contribute to the stabilization of climate change shocks, while programs tagged as adaptation strategies are intended to reduce the vulnerabilities to climate change shocks. Implicitly, OECD (2011) definitions imply a long term gestation period for adaptation programs, given the possibility of a time lag between current climate change shock and evolving adverse impacts in the future (Fazey, et al., 2010; Cooper, et al., 2013). The data on adaptation programs was collected sequentially. First, the authors highlighted all programs in the budgets that were directly linked to coping with climate change shocks. Thereafter, the authors separated adaptation programs by following the OECD (2011) definitions. Table 4 provides a summary of the data collected in both sequences.

Table 4. Selected climate change mitigation and adaptation programs in the Nigerian budget (2013 – 2020)

Year	Number of Climate Change Programs	Number of Mitigation Programs	Number of Adaptation Programs
2013	560	281	279
2014	508	337	171
2015	582	341	241
2016	780	485	295
2017	524	335	189
2018	671	458	213
2019	274	128	146
2020	282	126	156
Total	4181	2491	1690

Source: Authors' Compilation from 2013 – 2020 Nigerian Budgets.

Figure 2. Shares of unadjusted adaptation and mitigation programs to total climate change programs in the Nigerian budget (2013 – 2020)
Source: Authors' Compilation from 2013 – 2020 Nigerian Budgets

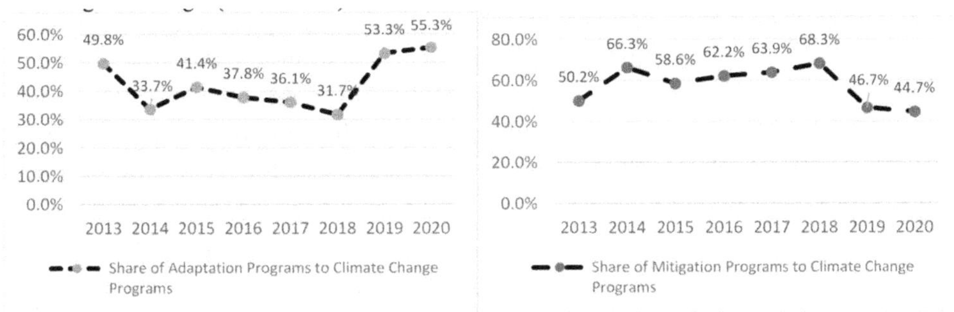

Figure 3. Shares of adjusted adaptation and mitigation programs to total climate change programs in the Nigerian budget (2013 – 2020)
Source: Authors' Compilation from 2013 – 2020 Nigerian Budgets

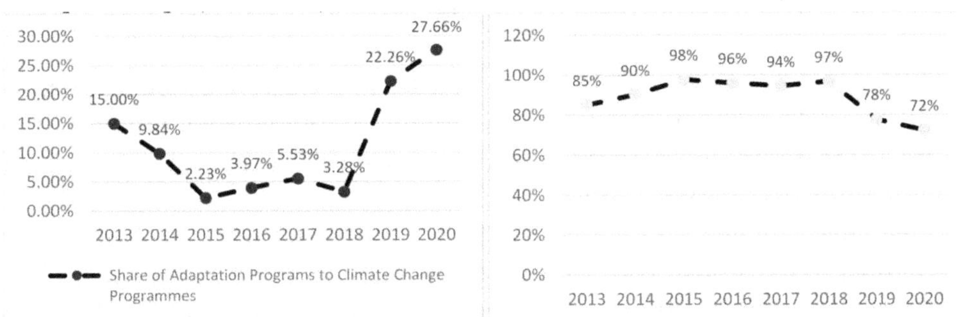

Table 4 indicates a relative preference for mitigation programs over adaptation programs, although this preference has begun to change from 2019. For each budget year, the number of mitigation programs surpassed the number of adaptation programs, except in 2019 and 2020. Figure 3 highlights the comparative differences between the federal government's choices of mitigation over adaptation. A closer look at the type of adaptation programs in respective year budgets indicate that a majority of adaptation programs are either flood control, erosion control or irrigation projects. When controlled for these three types of programs, the share of adaptation programs to all climate change efforts significantly reduces, compared to Mitigation Programs (see Figure 3). When scaled with the number of climate change programs, the share of mitigation programs consistently remained above 70% of total climate change programs. In contrast, the share of adaptation programs fell from 15% in 2013 to 2.23% in 2015, but rebounded moderately to 27.66% as at 2020. This indicates the overwhelming preference for flood control, erosion control and irrigation projects as core adaptation strategies. A plausible reason for the choice of these programs can be attributed to the susceptibility of Nigeria's rain fed agricultural output and yield to adverse climate change shocks (Hider, 2019), and a commitment to reducing this vulnerability as stated in Nigeria's third communication to the UNFCCC (Federal Ministry of Environment, 2020).

The rationale for adaptation programs arises from the fact that the adverse effects of climate change shocks are time bounded. The effects of these shocks often become visible in the future, which necessitate the need for longer term corrective measures. However, due to office term limits of decision makers and public servants – mostly four years and in the short term – decisions regarding how to deal with climate change shocks (choice to mitigate or adapt) would often depend on the expected tenure of the decision maker (Adger 2006). In Nigeria, there is documented evidence of the high turnover in expected tenure limits for appointed and elected decision makers in the bureaucracy (Fashagba, 2009; Omotola, 2010: Olorunmola, 2016), which limits the options of political decision makers to favour choices with expected immediate impacts over choices with longer term impacts (De Mesquita, 2002; Urwin and Jordan, 2008).

The other contributory reason for the preference for mitigation programs is the selection of constituency programs by members of the parliament. The members of the Nigerian parliament are by law allowed to delegate constituency projects to bureaucrats for implementation. Rogger (2014) notes that the choice of constituency projects among parliamentarians in Nigeria are mostly influenced by the degree of political competition and re-election risks. As such, regarding the adverse impacts of climate change, parliamentarians often lean towards projects that mitigate the immediate concerns of the electorates, especially when there is a high risk of losing elections. This preference for programs that deal with immediate climate change challenges ensures that constituency projects become more of mitigation rather than adaptation measures. For example, when mitigation programs are categorized into the provision of boreholes and street lightning, each of these two categories account for 863 and 549 programs respectively. When combined, the two programs make up approximately 83% of all adaptation programs but 56% of mitigation programs in the budgets from 2013 to 2020. Apart from the conspicuous dominance of these program categories in the budget, the choice of these categories is premised on the documented impact of climate change on access to water for farming and livelihood, the adverse effects of erosion of farm harvests, and limited access to electricity (NIMET, 2008).

Table 5. Number of mitigation programs by categories (2013 – 2020)

Budget Year	Boreholes	Street Lights
2013	193	32
2014	33	63
2015	72	35
2016	241	45
2017	121	147
2018	141	226
2019	24	0
2020	38	1
Total	863	549

Source: Authors' Compilation from 2013 – 2020 Nigerian Budgets

Although the data suggests the federal government's preference for mitigation programs over adaptation programs, there is ample evidence from budget data that the government remained committed to funding adaptation programs. However, the question therefore is what is the nature of these adaptation projects and how do they align to projects stipulated in the NASPA – CCN policy? Given the dominance of irrigation projects and erosion and flood control programs in the budget, the study presents an unadjusted and adjusted findings from employing the Climate Budget Tagging tool in two ways. The weighted scores presented in the unadjusted CBT, recognizes all the adaptation programs in the budget. On the other hand, in the adjusted CBT, the study neglects the scoring of irrigation projects, erosion control and flood control in the budgets. The choice of an adjusted and unadjusted CBT tool is premised first on the overriding influence of the three programs on the number of identified climate change adaptation programs. Restricting the analysis of the CBT scoring to an unadjusted data, may also exclude possible insights regarding other adaptation programs. In addition, Nigeria's communication to the UNFCCC is clear on the significant objective regarding the choice of these programs - to reduce vulnerabilities in the agricultural sector, human settlement, livelihood and disaster. Secondly, according to OECD (2011), these programs do not automatically qualify as an adaptation program. There must be sufficient evidence from the objectives of the programs to justify that the activities associated with flood control, erosion control and irrigation projects, contribute significantly to climate change adaptation.

The CBT scores by year and adaptation program, for both the unadjusted and adjusted budget data are presented in Figure 5. For the yearly CBT scores for the unadjusted budget data, the trend in scores show a continuous decline. This implies that the cumulative commitment towards principal and significant policy objectives towards climate change adaptation is on the decline. However, when considering the declining number of flood control, erosion control and irrigation projects (FEI) in the adjusted data (figure 4), there is a change in the trajectory of policy commitment, especially from 2019. The increase in the CBT score in 2019 and 2020, irrespective of the dominance of the FEI programs, underscores the government's intent on accommodating other types of adaptation programs. In the unadjusted budget data that accounts for the FEI type programs, Agriculture, Human Settlement and Housing, Livelihood, and Disaster, Migration and Security had the highest CBT scores. However, when the study de-emphasises FEI type programs, the targeted sectors benefiting from this more accommodating adaptation programs expanded – Agriculture, Health and Sanitation, Human Settlement and Housing, Energy, Livelihood, Disaster, Migration, and Security, and Vulnerable Groups (Figure 6).

Figure 4. Number of new and ongoing (flood, erosion and irrigation) FEI adaptation programs in the budget (2013 – 2020)
Source: Authors' Compilation form 2013 – 2020 Nigerian Budgets

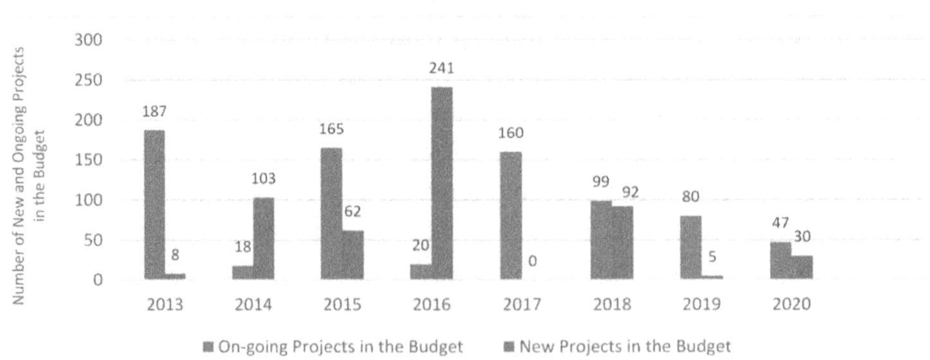

The declining number of FEI projects in the budgets as at 2015 is consistent with the current governments' policy stance towards the completion of all ongoing projects[2]. There is ample evidence in the literature on the abandonment of capital projects and policy discontinuity in Nigeria, especially when there is a change in governance (Williams, 2017; Ihuah and Benebo, 2014). The antecedence of such policy discontinuity has been the concentration of new project over ongoing projects in the budget. Data from the CBT scoring tool shows that the number of new climate change adaptation programs are declining, with the government focusing on completing ongoing programs.

The aggregate score by type of adaptation program gives an insight of the government's prioritization of each adaptation sector. In figure 5 the study presents both the adjusted and unadjusted CBT scores for each sector as stipulated in the NASPA – CCN policy. As expected, due to the significant number of flood control, erosion control and irrigation projects and the consequent impact on Agriculture, Human Settlement and Housing, Livelihood, and Disaster, Migration and Security, the unadjusted data indicates that climate change adaptation programs in the budget from 2013 to 2020 had prioritized these sectors. Conversely, the adjusted data proffers a more insightful analysis. When controlling for the FEI type programs, the adjusted data highlights government's efforts towards education, livelihood, forestry, Biodiversity, and health and sanitation. In addition, the transport and communication sector as well as industry and commerce had the lowest scores. The conclusion from both the adjusted and unadjusted budget data is that, despite the concentration of adaption programs on flood control, erosion control and irrigation, the government has proactively de-emphasized these FEI programs by remaining committed to completing already ongoing FEI programs, while expanding the scope of adaptation programs that have both principal and significant impacts on other sectors. Despite this expansion in the scope of adaptation programs, industry, commerce, communication and transport sectors are the least prioritized in the budget from 2013 to 2020.

The low CBT scores for industry and commerce is attributed to the degree of industrialization in Nigeria, which is low and its consequent low GHG emission (Emodi, et al., 2017). Afsah, et al (201) recognizes the limitations of compliance to adaptation and regulatory strategies' to industrial pollution in developing countries like Nigeria, based on the need to first understand the preconditions for such regulations and adaptation strategies. The authors highlight substantial transaction costs, inadequate integrated information, low public mandate and low capacity to implement adaptation strategies regarding climate change shocks. For transport and communication infrastructure, adverse climate change shocks like increases in temperature and rainfall softens the roads, which leads to the damage of such infrastructure. This damage increases the costs of moving people, goods and services and the operations of other sectors that depend on transport and communication infrastructure.

Despite the associated importance of this infrastructure, the CBT scoring tool indicates a relatively low score. A possible reason for this is premised on the type of adaptation programs targeted at the sector. The NASPA – CCN specifically targets the development of alternative route, expansion of highways and improvements in road maintained as adaptation programs. However, following the established criteria for identifying adaptation programs and their objectives, the programs targeted at transportation and communication are very limited. For a majority of the programs which aligned with the NASPA – CCN proposed adaptation programs for transportation and communication, and were included in the budget from 2013 – 2020, their stated objectives (both principal and significant) did not align with the scoring criteria.

Figure 5. Unadjusted and adjusted aggregate CBT score by year and sector (2013 – 2020)
Source: Authors' Compilation form CBT and 2013 – 2020 Nigerian Budget.

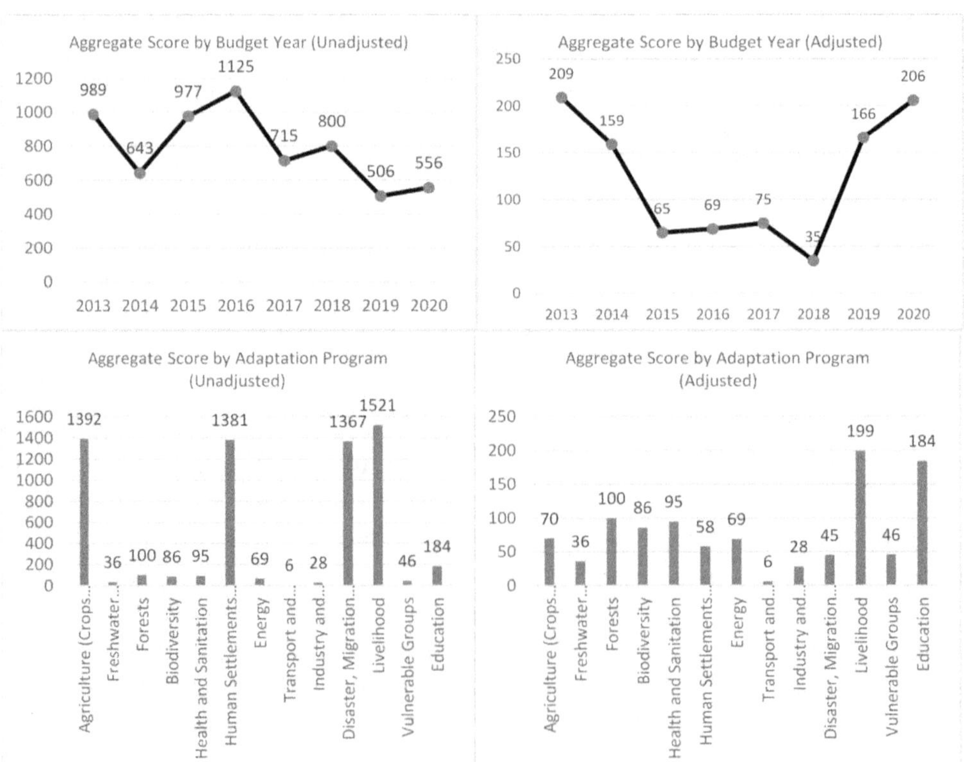

The study also presents a disaggregated view of the CBT scores for all sectors identified in the NASPA – CCN policy (Figure 6). For this analysis, the authors used the adjusted data given the overriding effects of the FEI type adaptation programs. Thus, the analysis provided do not include the possible impact of the FEI type adaptation programs on various sectors. However, a preliminary analysis of the unadjusted data indicates that. With the inclusion of the FEI type adaptation programs, only Agriculture, Human Settlements and Housing, Livelihood, and Disaster, Migration and Security are affected. The effects found were also very similar for each of these sectors. The disaggregated view provides a clearer picture of which sector the government prioritizes and possible areas of improvements.

Generally, the scores for most of the sectors were on a decline prior to 2018, with a reversal of the trend in 2019. The decline in the scores may not be disassociated from the government's stance on focusing on completing all ongoing projects. However, it is recognized that Nigeria was in a recession in 2015 and had moderate growth in 2016, which according to Obani and Gupta (2020), in recessions the general perception with regards government policy is to focus on economic recovery, with climate change framed as hampering economic recovery. The intuition is that during periods of recession and economic recovery, developing economies tend to favour programs that support economic growth rather than climate change programs. All sectors except livelihood and education had low scores for the period of 2015 and 2016. The focus on livelihood and education is consistent with the nature of adverse climate change shocks which predominantly affects livelihood (Amos, et al., 2015), most of which are vulnerable homes (Ebele and Emodi, 2015), which necessitates changing their practices to adapt effectively, through education (Amanchukwu, et al., 2015).

The upshot in adaptation programs from 2018 underscores the need to mitigate negative outcomes of recent climate change shocks in Nigeria. Incidences of increased temperature, flooding, erosion, has promulgated increased occurrence of clashes between farmers and herders in North East Nigeria, low incomes for farming households and increased vulnerability to food security (Haider, 2019). Coupled with the drop in crude oil prices in 2019 and the current global health pandemic in 2020, the government through the Federal Ministry of Environment is committed to upscaling adaptation programs that reduce household vulnerabilities (Federal Ministry of Environment, 2020).

Figure 6. Disaggregated CBT score by year and sector (2013 – 2020)
Source: Authors' Compilation from CBT Scoring Tool

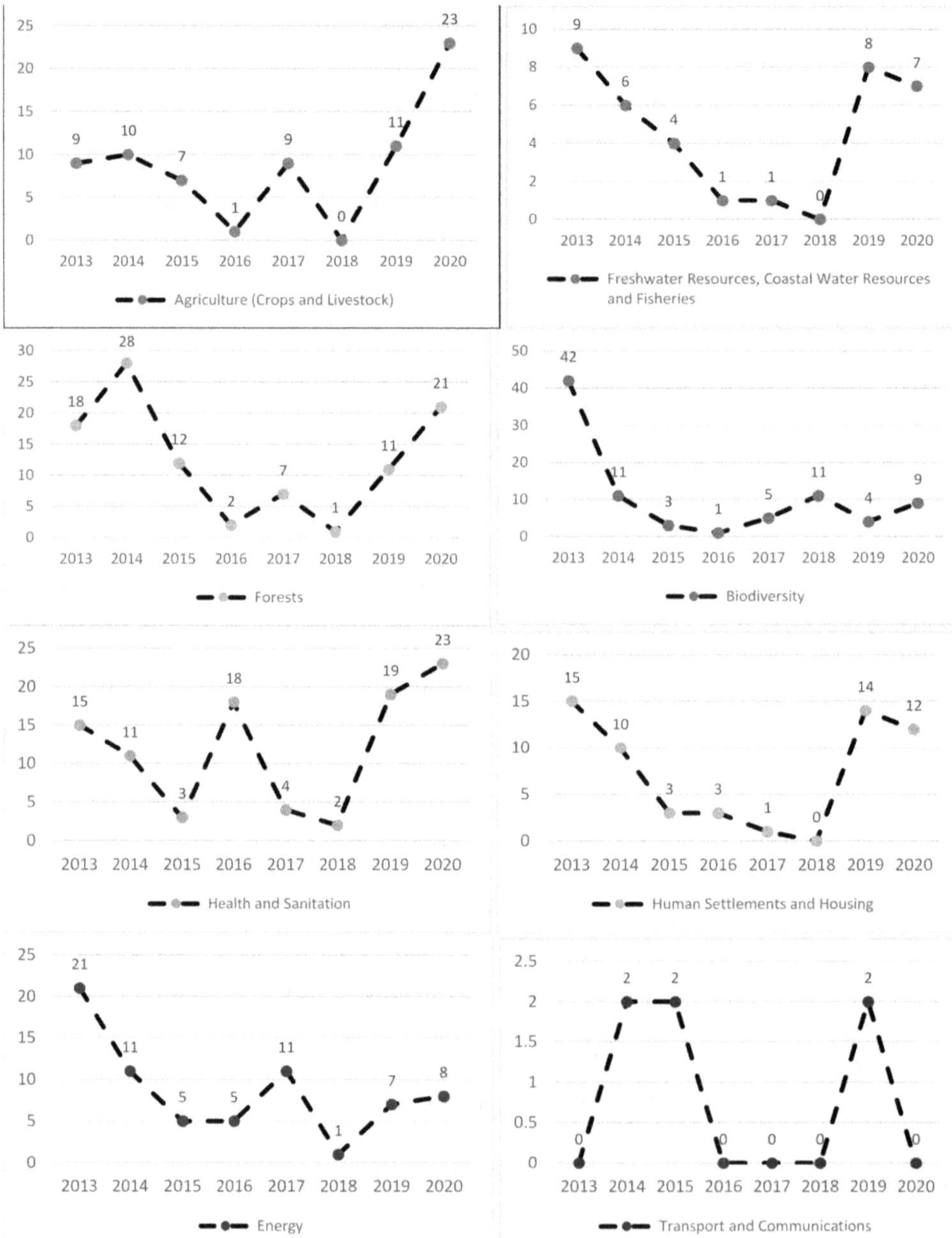

Another important question the paper intends to provide answers to, is the determination of any progressive achievements in the budgetary allocations to climate change adaptation programs. Progressive achievements with regards to climate change adaptation, focuses on budgetary expenditure on adaptation programs, with the aim of attaining the objectives of the NASPA – CCN policy. The authors first adjust budgetary allocations for inflation to capture the real monetary value of expenditure. A disaggregated expenditure by each priority sector would provide a clearer indication of the requirement of each sector. However, due to the use of the CBT scoring tool, a disaggregated approach may not be feasible. The major challenge is double counting of budgetary allocations for a program which may have impact on more than one priority sector. As such, the authors only provide the aggregated expenditures, adjust them for inflation.

Table 6 provides the aggregate expenditures for all adaptation programs for each year. The trend in allocations indicate a decline in budgetary allocations, despite accounting for a recession year in 2015 and expected recovery in 2016. However, despite the initial large allocation in 2013, the federal government has not been able to replicate this commitment in subsequent years. This is actually worrisome as it does not align with the provisions in the NASPA – CCN policy's priority implementation action of mobilizing resources. Another concern regarding the priority implementation with regards to mobilizing resources is the focus on international sources of funds for funding adaptation programs. Using Consumer Price Index (CPI), with 2012 as a base year, the budgetary allocation is adjusted to reflect the effects of inflation on purchasing power. Figure 7 indicates that growing inflating severely undermines the real purchasing power of budgetary allocations to adaptation programs. Although the degree of reduction in purchasing power is lower in 2013 to 2016, the gap between inflation adjusted and nominal budgetary allocations expanded significantly. The essence of adjusting for inflation underscores the extent to which more funds may need to be allocated to adaptation programs due to the eroding power of inflation. From figure 7, increasing inflation occasioned by adverse economic outcomes, erodes the financial commitments of the federal government towards prioritizing climate change adaptation programs.

Table 6. Aggregate budget expenditures for adaptation and yearly share of adaptation budget (2013 – 2020)

Year	Total (N)	Share of Total Adaptation Budget
2013	10,753,404,222	34.05%
2014	2,591,421,756	8.20%
2015	238,646,377	0.76%
2016	1,786,363,620	5.66%
2017	6,122,821,038	19.39%
2018	1,099,037,404	3.48%
2019	4,986,075,679	15.79%
2020	4,006,552,629	12.69%
Total Adaptation Budget (2013 – 2020)	31,584,322,725	

Source: Authors' Compilation from CBT Scoring Tool and Nigerian 2013 – 2020 Budgets.

Figure 7. Effect of inflation adjusted budget for adaptation expenditures (2013 – 2020)
Source: Authors' Compilation from CBT Scoring Tool.

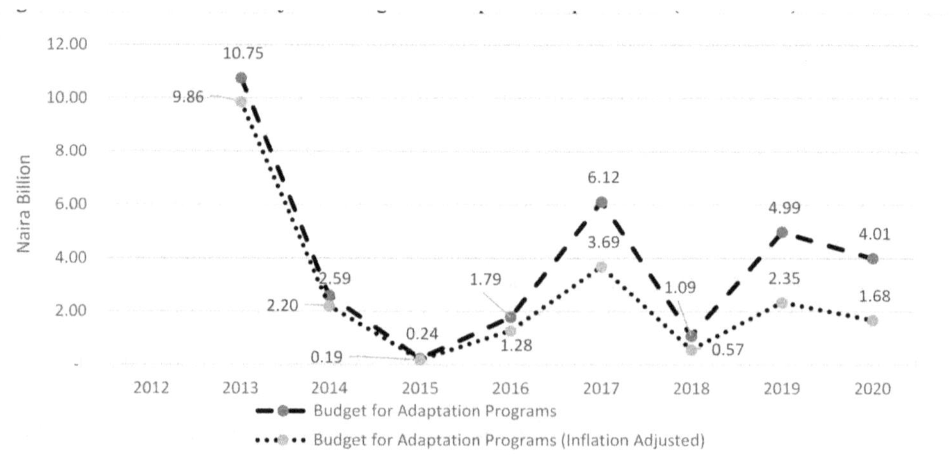

FUTURE RESEARCH DIRECTIONS

There exist some limitations in the application of the CBT in tracking both adaptation and mitigation financing for developing economies. These limitations are based on incomplete and most times, unavailable data. Also, for developing economies that rely on funding from international development partners, tracking such financing may be a big issue, especially as reporting and documentation standard may differ. As such, further research in adapting the CBT towards tracking the totality of climate change financing, domestic and foreign sources remains an important niche.

CONCLUSION

Taken together, the data and analysis of the Federal Government's financial commitment to its adaptation program indicate that, the Nigerian government's financial commit has changed in recent times, while some sectors as indicated in the NASPA – CCN policy are prioritized over others. A major finding in the paper is the overwhelming dominance of mitigation programs over adaptation programs. Despite this dominance, the paper also found that the type and structure of programs in the budget align with the objectives of the NASPA – CCN policy. This indicates government's policy consistence with implemented programs. A highlighted reason for this consistency is that a majority of Nigeria's climate change shocks are linked to increased temperatures and increased rainfall, which have significant adverse effects on agriculture and livelihood. The authors recommend that government prioritize both mitigation and adaptation strategies concurrently, and not neglect other sectors (apart from agriculture and livelihood) which also suffer significantly from climate change shocks.

The study also found that specific sectors – transport, telecommunication, industry and commerce are severely lacking in adaptation programs. We noted that a possible reason for this is the small contributions of the manufacturing sector to GDP as well as the slow pace of industrialization in the country. As such, the government does not prioritize the negative externalities that arise from the process of industrialization. However, Nigeria is on the process of becoming a highly industrialized country, given current government economic policy stance. Thus, the study recommend that Nigeria's industrialization and commercialization policies (short and long term) should include strategies and programs that are geared towards expected increases in climate change shocks. To remain consistent with the objectives of the NASPA – CCN policy, all future development plans should incorporate international best practices that reduce emissions, mitigate the negative externalities from increased production, and proffer simultaneous adaptation strategies for future climate change shocks.

Government's financial commitment had been on the decline prior to 2018, but has been scaled upwards in 2019 and 2020. The decline in financial commitments in previous years was attributed to the governments' policy stance on completing all ongoing projects, the 2015 recession and a preference for growth stimulating projects over climate change adaptation programs. With regards to investigating if there are progressive achievements in the financing of adaptation programs in the budget, the study did not find substantial evidence from the CBT scores and inflation adjusted allocations to support the assertion of progressive achievements. The changes in yearly budgets to adaptation programs reflects a commitment towards completing ongoing adaptation projects, but does not align with the provisions in the NASPA – CCN policy, which mandates consistent increase in financial resources to accommodate adaptation programs. A highlighted reason for this is the negative effects of crude oil price volatilities in the international market and the Nigeria's susceptibility to external shocks. Thus, whenever revenue projections are not met, the government prioritizes growth recovery programs over programs associated with mitigating or adapting to climate change shocks. The study recommend that during periods of recession or reduced revenues, the government could lean towards international sources of climate change financing rather than depend of domestic revenues.

ACKNOWLEDGMENT

This research was supported by the National Institute for Legislative and Democratic Studies.

REFERENCES

Adger, W. N. (2006). Vulnerability. *Global Environmental Change*, *16*(3), 268–28. doi:10.1016/j.gloenvcha.2006.02.006

Afsah, S., Laplante, B., & Wheeler, D. (1996). Controlling Industrial Pollution: a new paradigm. *World Bank policy research working paper*, (1672). World Bank.

Amanchukwu, R. N., Amadi-Ali, T. G., & Ololube, N. P. (2015). Climate change education in Nigeria: The role of curriculum review. *Education*, *5*(3), 71–79.

Ampaire, E. L., Jassogne, L., Providence, H., Acosta, M., Twyman, J., Winowiecki, L., & Van Asten, P. (2017). Institutional challenges to climate change adaptation: A case study on policy action gaps in Uganda. *Environmental Science & Policy*, *75*, 81–90. doi:10.1016/j.envsci.2017.05.013

Ayers, J. (2009). International funding to support urban adaptation to climate change. *Environment and Urbanization*, *21*(1), 225–240. doi:10.1177/0956247809103021

Barr, R., Fankhauser, S., & Hamilton, K. (2010). Adaptation investments: A resource allocation framework. *Mitigation and Adaptation Strategies for Global Change*, *15*(8), 843–858. doi:10.100711027-010-9242-1

Bouwer, L. M., & Aerts, J. C. (2006). Financing climate change adaptation. *Disasters*, *30*(1), 49–63. doi:10.1111/j.1467-9523.2006.00306.x PMID:16512861

Bowen, A., Cochrane, S., & Fankhauser, S. (2012). Climate change, adaptation and economic growth. *Climatic Change*, *113*(2), 95–106. doi:10.100710584-011-0346-8

Brooks, N., Anderson, S., Burton, I., Fisher, S., Rai, N., & Tellam, I. (2013). *An operational framework for tracking adaptation and measuring development*. TAMD.

Buckley, D. (2014). Mobilising Adaptation Finance: The status of public finance related to national funding for developing countries. Presented for the Second Forum of the Standing Committee on Finance 21 June 2014. UNFCCC. https://unfccc. int/sites/default/files/s3_1_daniel_b_scf_adaptation_finance_seminar_undp_dbuckley.pdf

Burmeister, H., Cochu, A., Hausotter, T., & Stahr, C. (2019, October 21). *Financing adaptation to climate change – an introduction (Adaptation Briefings)*. Adaptation Community. https://www.adaptationcommunity.net/wp-content/uploads/2019/10/2019-10_adelphi_Adaptation-Briefings_Financing-Adaptation_an-Introduction.pdf

Challinor, A., Wheeler, T., Garforth, C., Craufurd, P., & Kassam, A. (2007). Assessing the vulnerability of food crop systems in Africa to climate change. *Climatic Change*, *83*(3), 381–399. doi:10.100710584-007-9249-0

Cooper, P. J. M., Stern, R. D., Noguer, M., & Gathenya, J. M. (2013). Climate change adaptation strategies in Sub-Saharan Africa: foundations for the future. In Climate change—realities, impacts over ice cap, sea level and risks. Intech Open. doi:10.5772/55133

CPI. (2017). *Global Landscape of Climate Finance 2017.* CPI. https://climatepolicyinitiative.org/wpcontent/uploads/2017/10/2017-Global-Landscape-of-Climate-Finance.pdf

CPI. (2018). *Global Climate Finance: An Updated View 2018.* CPI. https://climatepolicyinitiative.org/wp-content/uploads/2018/11/Global-Climate-Finance-AnUpdated-View-2018.pdf

CPI. (2018a). *Understanding and Increasing Finance for Climate Adaptation in Developing Countries.* CPI. https://climatepolicyinitiative.org/wp-content/uploads/2018/12/Understanding-and-IncreasingFinance-for-Climate-Adaptation-in-Developing-Countries-1.pdf

De Mesquita, B. B., Morrow, J. D., Siverson, R. M., & Smith, A. (2002). Political institutions, policy choice and the survival of leaders. *British Journal of Political Science*, *32*(4), 559–590. doi:10.1017/S0007123402000236

Dodman, D., Ayers, J., & Huq, S. (2009). *Building Resilience in 2009 State of the World; into a Warming World.* The Worldwatch Institute.

Ebele, N. E., & Emodi, N. V. (2016). Climate change and its impact in Nigerian economy. *Journal of Scientific Research and Reports*, *10*(6), 1–13. doi:10.9734/JSRR/2016/25162

Emodi, N. V., Emodi, C. C., Murthy, G. P., & Emodi, A. S. A. (2017). Energy policy for low carbon development in Nigeria: A LEAP model application. *Renewable & Sustainable Energy Reviews*, *68*, 247–261. doi:10.1016/j.rser.2016.09.118

European Bank for Reconstruction and Development. (2014, September). *Joint Report on MDB climate Finance.* EBRD. https://www.ebrd.com/downloads/news/mdb-climate-finance-2013.pdf

Fashagba, J. Y. (2009). Legislative Oversight under the Nigerian Presidential System. *Journal of Legislative Studies*, *15*(4), 439–459. doi:10.1080/13572330903302497

Fazey, I., Gamarra, J. G., Fischer, J., Reed, M. S., Stringer, L. C., & Christie, M. (2010). Adaptation strategies for reducing vulnerability to future environmental change. *Frontiers in Ecology and the Environment*, *8*(8), 414–422. doi:10.1890/080215

Federal Ministry of Environment. (2020). *Third National Communication (TNC) of the Federal Republic of Nigeria: United Nations Framework Convention on Climate Change* (UNFCCC). FME. https://www4.unfccc.int/sites/SubmissionsStaging/NationalReports/Documents/187563_Nigeria-NC3-1-TNC%20NIGERIA%20-%2018-04-2020%20-%20FINAL.pdf

Ford, J. D., Berrang-Ford, L., Biesbroek, R., Araos, M., Austin, S. E., & Lesnikowski, A. (2015). Adaptation tracking for a post-2015 climate agreement. *Nature Climate Change*, *5*(11), 967–969. doi:10.1038/nclimate2744

Haider, H. (2019). Climate change in Nigeria: Impacts and responses. K4D Helpdesk Report 675. Brighton, UK: Institute of Development Studies.

Ihuah, P. W., & Benebo, A. M. (2014). An assessment of the causes and effects of abandonment of development projects on real property values in Nigeria. *International Journal of Research in Applied, Natural and social sciences, 2*(5), 25-36.

Le, H., & Baboyan, K. (2015). *Climate Budget Tagging County – Driven Initiative in Tracking Climate Change Expenditure: The Case of Bangladesh, Indonesia, Nepal and Philippines*. (UNDP Working Paper). UNDP.

Lim, B., Spanger-Siegfried, E., Burton, I., Malone, E., & Huq, S. (2005). *Adaptation policy frameworks for climate change: developing strategies, policies and measures*. United Nations Development Programme.

Lobell, D. B. (2014). Climate change adaptation in crop production: Beware of illusions. *Global Food Security*, *3*(2), 72–76. doi:10.1016/j.gfs.2014.05.002

Lobell, D. B., Burke, M. B., Tebaldi, C., Mastrandrea, M. D., Falcon, W. P., & Naylor, R. L. (2008). Prioritizing climate change adaptation needs for food security in 2030. *Science*, *319*(5863), 607–610. doi:10.1126cience.1152339 PMID:18239122

Micale, V., Tonkonogy, B., & Mazza, F. (2018). *Understanding and Increasing Finance for Climate Adaptation in Developing Countries*. Climate Policy Initiatve.

Moser, S. C., & Ekstrom, J. A. (2010). A framework to diagnose barriers to climate change adaptation. *Proceedings of the National Academy of Sciences of the United States of America*, *107*(51), 22026–22031. doi:10.1073/pnas.1007887107 PMID:21135232

Mukheibir, P., Kuruppu, N., Gero, A., & Herriman, J. (2013). Overcoming cross-scale challenges to climate change adaptation for local government: A focus on Australia. *Climatic Change*, *121*(2), 271–283. doi:10.100710584-013-0880-7

New, M., Liverman, D., Schroder, H., & Anderson, K. (2011). Four degrees and beyond: the potential for a global temperature increase of four degrees and its implications: Introduction. *Philosophical Transactions of the Royal Society A: Mathematical, Physical and Engineering Sciences, 369*(1934), 6-19.

NIMET (2008). *Nigeria Climate Review Bulletin 2007*. Nigerian Meteorological Agency.

Obani, P. C., & Gupta, J. (2016). The impact of economic recession on climate change: Eight trends. *Climate and Development*, *8*(3), 211–223. doi:10.1080/17565529.2015.1034226

ODI. (2017). *Climate Finance Thematic Briefing: Adaptation Finance*. ODI. https://www.odi.org/sites/odi.org.uk/files/resource-documents/12073.pdf

OECD. DAC. (2011). Handbook on the OECD-DAC Climate markers. *Paris: Organisation for Economic Co-operation and Development's Development Assistance Committee (DAC)*. OECD. https://www. oecd. org/dac/stats/48785310. pdf

Olorunmola, A. (2016). *Cost of politics in Nigeria*. Westminster Foundation for Democracy.

Omotola, J. S. (2010). Elections and democratic transition in Nigeria under the Fourth Republic. *African Affairs*, *109*(437), 535–553. doi:10.1093/afraf/adq040

Organization for Economic Cooperation and Development. (2011, September). *Handbook on the OECD - DAC Climate Markers*. OECD. https://www.oecd.org/dac/stats/48785310.pdf

Resch, E., Allan, S., Álvarez, L. G., & Bisht, H. (2017). *Mainstreaming, accessing and institutionalising finance for climate change adaptation. ACT Learning Paper*. OPM.

Rogger, D. (2014). The causes and consequences of political interference in bureaucratic decision making: Evidence from Nigeria. *Job Market Paper, 12*(1), 1–22.

Schuftan, C. (2005). Dignity Counts: A guide to using budget analysis to advance human rights. *Social Change, 35*(1), 143–146. doi:10.1177/004908570503500113

Smit, B., & Skinner, M. W. (2002). Adaptation options in agriculture to climate change: A typology. *Mitigation and Adaptation Strategies for Global Change, 7*(1), 85–114. doi:10.1023/A:1015862228270

Stafford, M. S., Horrocks, L., Harvey, A., & Hamilton, C. (2011). Rethinking adaptation for a 4° C world. *Philosophical transactions. Series A, Mathematical, physical, and engineering sciences, 369*(1934), 196-216.

Thorarinsdottir, T. L., & de Bruin, K. (2016), Challenges of climate change adaptation, *Eos, 97,* https://doi.org/ doi:10.1029/2016EO062121

UNDP. (2012). *Climate Public Expenditure and Institutional Reviews (CPEIRs) in the Asia-Pacific Region - What have We Learnt?* UNDP. https://www.asia-pacific.undp.org/content/dam/rbap/docs/Research%20&%20Publications/democratic_governance/APRC-DG-2012-CPEIRLessonsLearnt.pdf

UNDP. (2016a). *Charting New Territory: A stock take of climate change finance frameworks in Asia-Pacific.* UNDP. https://www.climatefinance-developmenteffectiveness.org/sites/default/files/documents/09_06_16/Charting%20New%20Territory%20%20A%20Stocktake%20of%20Climate%20Change%20Financing%20Frameworks%20in%20Asia%20Pacific.pdf

UNDP. (2016b). *The Adaptation Finance Gap Report.* UNDP. orbit.dtu.dk/ws/files/177810752/50313_UNEP_GAP_report_2016_v5_SB.pdf

UNFCCC. (2018). *UNFCCC Standing Committee on Finance 2018 Biennial Assessment and Overview of Climate Finance Flows.* UNFCCC. https://unfccc.int/sites/default/files/resource/2018%20BA%20Technical%20Report%20Final.pdf

UNFPA. (2009). *Financing that Makes a Difference.* UNFPA. https://www.unfpa.org/sites/default/files/pub-pdf/climateconnections_5_finance.pdf

Urwin, K., & Jordan, A. (2008). Does public policy support or undermine climate change adaptation? Exploring policy interplay across different scales of governance. *Global Environmental Change*, *18*(1), 180–191. doi:10.1016/j.gloenvcha.2007.08.002

Vogel, B., & Henstra, D. (2015). Studying local climate adaptation: A heuristic research framework for comparative policy analysis. *Global Environmental Change*, *31*, 110–120. doi:10.1016/j.gloenvcha.2015.01.001

Williams, M. J. (2017). The political economy of unfinished development projects: Corruption, clientelism, or collective choice? *The American Political Science Review*, *111*(4), 705–723. doi:10.1017/S0003055417000351

World Bank. (2010). *World Development Report 2010. Development and Climate Change*. World Bank.

Ziervogel, G., Johnston, P., Matthew, M., & Mukheibir, P. (2010). Using climate information for supporting climate change adaptation in water resource management in South Africa. *Climatic Change*, *103*(3-4), 537–554. doi:10.100710584-009-9771-3

ADDITIONAL READING

Barnett, J. (2020). Global environmental change II: Political economies of vulnerability to climate change. *Progress in Human Geography*, *44*(6), 1172–1184. doi:10.1177/0309132519898254

Doshi, D., & Garschagen, M. (2020). Understanding Adaptation Finance Allocation: Which Factors Enable or Constrain Vulnerable Countries to Access Funding? *Sustainability (Basel)*, *12*(10), 4308. doi:10.3390u12104308

Hong, H., Karolyi, G. A., & Scheinkman, J. A. (2020). Climate finance. *Review of Financial Studies*, *33*(3), 1011–1023. doi:10.1093/rfs/hhz146

Mutiara, Z. Z., Krishnadianty, D., Setiawan, B., & Haryanto, J. T. (2021). Climate Budget Tagging: Amplifying Sub-National Government's Role in Climate Planning and Financing in Indonesia. In *Climate Change Research, Policy and Actions in Indonesia* (pp. 265–280). Springer. doi:10.1007/978-3-030-55536-8_13

Pardoe, J., Vincent, K., Conway, D., Archer, E., Dougill, A. J., Mkwambisi, D., & Tembo-Nhlema, D. (2020). Evolution of national climate adaptation agendas in Malawi, Tanzania and Zambia: The role of national leadership and international donors. *Regional Environmental Change*, *20*(4), 1–16. doi:10.100710113-020-01693-8

Schulz, K., & Feist, M. (2021). Leveraging blockchain technology for innovative climate finance under the Green Climate Fund. *Earth System Governance*, *7*, 100084. doi:10.1016/j.esg.2020.100084

Sheriffdeen, M., Nurrochmat, D. R., Perdinan, P., & Di Gregorio, M. (2020). Indicators to Evaluate the Institutional Effectiveness of National Climate Financing Mechanisms. *Forest and Society*, 358-378.

World Bank. (2020). *Mobilizing Financing for Climate Smart Investments in the Mekong Delta: An Options Note*. World Bank.

KEY TERMS AND DEFINITIONS

Adaptation: Any activity that reduces the vulnerability of human or natural systems to the impacts of climate change and climate-related risks, by maintaining or increasing adaptive capacity and resilience.

Climate Change: Refers to the increase in global warming and long term shifts in weather due to human activities.

Mitigation: Any Activity that reduces the possible long-term impacts of climate change on human life and the eco-system, through the reduction in emissions of greenhouse gasses.

ENDNOTES

1 The Budget analysis framework to be used is based on the guidelines of Dignity Counts. See Fundar – Centro de Análisis e Investigación, International Human Rights Internship Program and International Budget Project (2004). *Dignity Counts: A Guide to Using Budget Analysis to Advance Human Rights.* Fundar

2 The current administration first got elected in 2015, with the primary focus on completing all ongoing projects. This policy stance was stated in the 2016 Budget Speech and reiterated in subsequent budget speeches.

Chapter 8
ICT and Renewable Energy:
A Way Forward to Secure New Generation

Ahmad Tasnim Siddiqui
https://orcid.org/0000-0002-1884-9331
Babu Banarasi Das University, Lucknow, India

Nupur Soni
Babu Banarasi Das University, Lucknow, India

ABSTRACT

In the last few decades, we have seen remarkable growth in the technology sector. But this growth has cost us in the form of global warming and climate change. Information and communication technology (ICT) is the quickest grown area in the technology sector and impacted a lot on every part of our lives and is somehow responsible for global warming and climate change in the form of energy emissions. The increasing level of emission of carbon dioxide (CO2) and other gases from various sources has has great affect. ICT, RES, and strict governance can help to mitigate the problem and contribute to energy efficiency by becoming smart. It can contribute to developing smart homes, smart grids, smart logistics, and other smart devices. Using renewable energy sources (RES), small and medium scale systems, computers and related peripherals, and electric and solar vehicles may contribute to conserving the environment and mitigate climate change and global warming. If we can work honestly, we can secure our future generations by making technologies smarter by integrating RES.

DOI: 10.4018/978-1-7998-8638-9.ch008

1. INTRODUCTION

The key driver of socioeconomic growth in most economies is technological development. As ICT advances, countries are able to develop economically as well as improve the quality of life for people. With the advent of ICT, communication methods around the world have changed, and now the world is a "global village" (Zeeshan et. *al*., 2022). There is growing concern over human activities harming the environment, especially climate change. The issue of climate change poses serious challenges to people around the world. As a result, citizens are more in danger, economic gains will be undermined, and social and economic development will be hampered. It also threatens to worsen access to basic services and the quality of life for citizens. As a result, countries must adapt to climate change to survive. Climate change adaptation can be greatly enhanced by information and communication technologies. Climate change adaptation strategies can be enriched and advanced with it. It is difficult to develop and implement climate change adaptation strategies in urban contexts because of the complexity of the environment. However, it also enables the ICT sector to play a greater role in adapting to climate change in countries (ITU, 2015).

For the technologies reliant on systems, electrical appliances use a large amount of energy. Nowadays, computing power is more than compared to the time when man was sent to the moon (Ahmed, F., et. al, 2016). Due to the technology-enabled lifestyle, we are consuming a lot of energy and the result is energy demand is expected to rise by 37% by 2035 on average of 1.4% a year (BP energy outlook 2035).

A constant increase in ''carbon footprint'' has also been caused by the revolution of ICT in a daily average life. In addition, to reduce energy consumption, raise environmental awareness, communicate effectively about environmental issues, and monitor and restore natural ecosystem functions, energy-efficient Informatics need to play an important role. Weather and geographical circumstances affect renewable energy sources and they are often uneven. ICT has taken the responsibility and initiatives have been taken. Globally, ICT plays a major role in combating climate change and protecting the environment in recent years. There is no doubt that climate change has an impact on ecosystems, water and food supplies, public health, agriculture, and infrastructure. In addition to increasing temperatures and sea levels, flooding and storms are becoming more frequent occurrences as well (Zacharoula, S. A., 2012). With the global climate change and instability of energy markets, ICT has begun to recognize the role it can play in designing energy-efficient and environmentally friendly technology and systems (Bronk, C. et. al, 2010).

Philip Song, CMO, Carrier BG, Huawei, in a keynote speech at Mobile World Congress (MWC) 2022, mentioned that by 2030, ICT technologies can reduce global CO2 emissions by 20 percent, which is ten times the carbon releases of the ICT industry itself, and effectively decouple economic growth from emissions growth. Huawei calls it "ICT enablement carbon handprints" (Baburajan, R., 2022). Over the last decade, network technology has changed a lot. It has traveled from normal cellular technology to 4G/LTE via 2G and now it is 5G. Network technology is always refining energy efficiency at every stage. As per the data revealed by Huawei, if we can successfully migrate 10% of 3G users to 4G/5G connectivity, we can achieve the 2% of energy-saving (Baburajan, R., 2022).

2. BACKGROUND AND LITERATURE REVIEW

Several studies have been done on the relationship between ICT, economic development, and other essential variables like a financial improvement, education, energy utilization, energy security, electricity usage, and social capital (Pradhan et. al (2015); Jorgenson et. al (2016); Erumban, A. A. & Kusum, D. (2016); Pfeifer et. al (2018); Céspedes-lorente, J. J., & Magán-díaz, A. (2018); Ribeiro et. al (2018); Abu, T. & Bressler, L. (2019); Pradhan et. al (2018); Hache, E. (2018); Yang et. al (2018); Thompson, M. (2018); Benighaus, C. (2019); Lin, B. & Chen, Y. (2019). These literature works have appeared with inconclusive results largely due to modifications in variable measurements, methodologies, territories of study, sample sizes, and level of development.

Economic growth has been greatly influenced by information technology, digitalization, and blockchain technologies during the globalization period (Saberi et al., 2019; Oliveira et al., 2020; Borowski, 2021). A number of studies have shown that the use of digitalization and ICT increases employment and reduces poverty (Coleman, 2005; Rot et al., 2020). The implementation of ICT and digitization increases the efficiency and security of the energy system (Garca-Quismondo et al., 2013; Mohsin Rahman & Mezbah-ul-Islam, 2012; K. Wang et al., 2018).

As a matter of fact, the significant evolutions of energy-related operations that occurred in the last decades address the generation aspects of the energy chain, the consumption behavior, and the market speculation, as well as the possibilities of setting up brand new services that rely upon economic. The term climate change refers to an overall change in the climate system over time (Glossary, 2001). Sometimes, the term specifically refers to changes in climate caused by human-caused activity rather than changes in climate that may have occurred naturally (Glossary, 2001). Furthermore, Human activity modifies the configuration of the global atmosphere, in addition to natural climate inconsistency over a comparable time period (Wikipedia). Carbon dioxide is formed when fossil fuels are burned in the lower atmosphere, causing global climate change. A 30 percent increase in atmospheric greenhouse gas concentrations since preindustrial times can be attributed to increased atmospheric carbon dioxide concentrations (The United Nations Framework Convention on Climate Change, 1994).

Researchers studied the impact of non-renewable and renewable energy on the environment. According to Riti et al. (2018), fossil fuel energy consumption increases emissions of greenhouse gases in both the short and long runs, while renewable energy consumption reduces emissions of greenhouse gases.

Global problems require global solutions, and climate change is one of the most serious challenges humanity faces in achieving sustainable development goals. In June 1992, at a conference in Rio de Janeiro, 155 countries signed the United Nations Framework Convention on Climate Change (The United Nations Framework Convention on Climate Change, 1994), pledging to reduce greenhouse gas emissions to enable sustainable economic development. During its March 2007 meeting, the European Council set three energy and climate change-related objectives: reducing greenhouse gas emissions by 20%, integrating renewable energy sources into the final energy mix in the EU by 20%, and reducing the use of primary energy in the EU by 20% by 2020. By using renewable energy and reducing energy consumption, Member States can reduce greenhouse gas emissions in 2020 and meet their goals. In response to climate change, the international community has held annual meetings of the parties to the Convention in various parts of the world. One such meeting was held in December 2011 in Durban, South Africa. It was specified that all "developed and developing nations will work for the very first time on a contract that should be officially obligatory, to be written by 2015 and to come into effect after 2020" (The Guardian Durban conference, 2011).

According to Dong et al. (2020), the mitigation benefits from renewable energy consumption outweigh the increased non-renewable energy consumption and economic growth. Increased renewable energy consumption will reduce CO2 emissions, but the demand and economic growth may mask the mitigation benefit.

With the advent of Information and Communication Technology, we are remaking the world in terms of enhancing access to information and human capital. This has an impact on the mental capacity of an individual, their ability to think, and their ability to communicate. Thanks to ICT (Azam et. al. 2021). As a result of the dissemination of education and health-related information, ICT can contribute to the development of people by removing impediments to their lives and improving their living standards (Palvia et al., 2015; Migliaccio 2016).

As part of the sustainable development goals (SDG), the United Nations insists that ambient air pollution be reduced globally, traditional energy is restricted from playing a larger role in the energy mix, and sustainable power sources must be supplemented as part of a more sustainable power mix (Azam A. et al., 2021). Azam et al (2021) also argue that for both developed and developing countries, information and communication technologies are vital to sustainable economic development.

3. CLIMATE CHANGE AND RENEWABLE ENERGY

Climate change normally occurs when a rise in temperature, wind, rainfall, and other elements start varying and continue for decades. But we are facing climate change due to human advancement in every area. It is human activity that is causing climate change and global warming. Over 75 percent of global greenhouse gas emissions come from fossil fuels, including coal, oil, and gas. CO2 emissions account for nearly 90% of global greenhouse gas emissions. Global warming and climate change are caused by the accumulation of greenhouse gas emissions on the surface of the Earth, trapping the sun's heat. The world is getting hotter very fast than before compared to history. An increase in temperature is causing drastic changes in the weather pattern. And due to this, nature's balance is affected. It is causing various risks to human beings and other lives on earth (Causes and effects of climate change).

Causes of climate change are Generating power, manufacturing goods, producing food, using transportation, cutting down forests, powering buildings, and consuming too much. And the adverse effects are More severe storms, a rise in temperatures, Increasing drought, Loss of species, More health risks, floods, etc. (Causes and effects of climate change).

Greenhouse gas (GHG) absorptions are at their highest levels in the last 2 million years and it continues to increase. Due to this, the earth is around 1.1°C hotter than the earth was in the 1800s. On record, the last decade was the warmest one. A report in 2018 shows that a number of scientists and governments agreed to mitigate the rise in global temperature to 1.5°C to avoid the worst climate effect and provide a lively climate globally. Switching energy systems from fossil systems to renewable energy systems like solar, wind, and hydropower will cause a decrease in emissions and control climate change (Climate Actions).

The majority of people in the world rely on fossil fuels. About 80 percent of the global population lives in countries that are net importers of fossil fuels, which means that around 6 billion people are dependent on fossil fuels from other countries, making them susceptible to geopolitical crises and shock. According to the approximation of The International Renewable Energy Agency (IRENA), 90 percent of the world's electrical energy can and should originate through renewable energy by 2050. Countries can expand their economies with renewable energy and prevent themselves from the price swings of fossil fuels, also they may grow their economies, create more jobs, and alleviate poverty with these practices (Renewable energy – powering a safer future).

Renewable energy is in fact the low-cost power option in most countries in the world. Renewable energy technologies are reducing their expenses swiftly. Between the years 2010 and 2020, the price of electricity production dropped by 85 percent. And the onshore price of wind energy fell by 56 percent while offshore wind energy fell by 48 percent. As per estimation, by 2030, 65% of the world's total cheap electrical energy could be from renewable sources. Carbon emissions could be massively reduced and climate change mitigated by decarbonizing 90 percent of the power sector by 2050 (Renewable energy – powering a safer future).

As per WHO, about 99% of the world's population breathes poor air quality and is at risk of diseases, and more than 13 million deaths are caused by improvable environmental factors, like air pollution. Fossil fuel combustion is the primary cause of unhealthy levels of fine particulate matter and nitrogen dioxide. Approximately $8 billion a day was lost to health and economic costs caused by air pollution from fossil fuels in 2018. By utilizing clean sources of energy, such as wind and solar, we can combat not only climate change, but also air pollution, health problems, and joblessness. By diversifying power sources, efficient, reliable renewable energy technologies can enhance resilience and energy security and make the system less vulnerable to market shocks (Renewable energy – powering a safer future).

China, the United States, and Brazil were the top countries for installing renewable energy in 2021. With a capacity of approximately 1,020 gigawatts, China ranked first in renewable energy installations. The United States ranked the second position, with a capacity of about 325 gigawatts.

Brazil got third place with 160 gigawatts and India secured the fourth position with 147 gigawatts. A key component of mitigating climate change is the use of renewable energy (Renewable energy capacity worldwide by country, 2020). Figure 1 given below shows the renewable energy capacity by country (Renewable energy capacity worldwide by country, 2020):

Figure 1. Renewable energy capacity 2021, by country

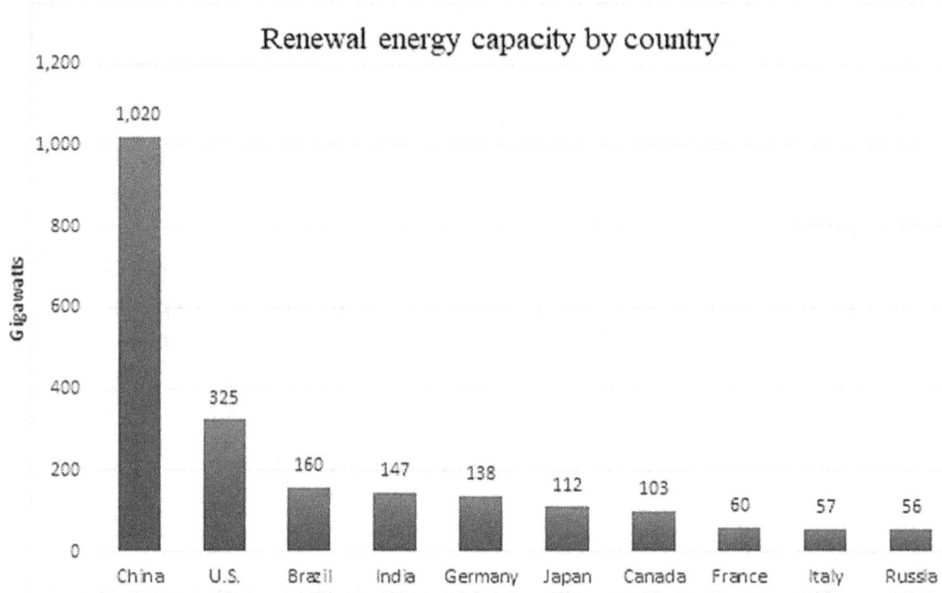

Climate Change and Renewable Energies in GCC Countries

The GCC (Gulf Cooperation Council) countries have experienced rapid and exceptional growth in every part of life due to their oil and natural gas returns. As a result, these countries have become the center of intense activities in many fields including geopolitics, military, economics, manufacturing, and construction, to name a few (Raouf, M., 2008).

It is not unusual for the GCC countries to face widespread, "traditional" environmental challenges. These include desertification, biodiversity loss, pollution in marine and coastal zones, air pollution, scarcity of drinking water, and poor quality of drinking water. There have been additional environmental problems that have arisen in recent years, especially those pertaining to military conflicts and construction debris. It is no surprise that climate change is one of the most pressing environmental challenges the region is facing (Raouf, M., 2008).

The rise in global temperatures is attributed to a wide variety of natural events, including rising sea levels, melting ice caps, and more severe storms.

As temperatures continue to rise, especially if they reach 2 to 3 degrees Celsius, the IPCC (Intergovernmental Panel on Climate Change) warns that massive species eliminations, widespread starvation, declining crop yields, and permanently rising sea levels could threaten major coastal cities. As a result of these developments, the GCC countries will be directly affected. As sea levels rise, natural and man-made islands around the world are likely to disappear. Bahrain may lose up to 15 kilometers of coastline if sea levels rise. Furthermore, there will be an increase in underground water salinity, more land degradation, and reductions in biodiversity on land and in the Gulf.

In comparison with catastrophic disasters such as hurricanes, tsunamis, and floods elsewhere in the world, Gulf coast ecological changes are minor. Gulf countries, however, are largely dependent on oil and gas export revenue, so the economic impact will be severer. Global shifts to renewable energy will cause serious problems for the Gulf region if they shift to alternative energy soon (Raouf, M., 2008).

Gulf countries are major petroleum exporters, but their use of fossil fuels has raised concerns about carbon emissions. They are under fire. The UAE and Saudi Arabia secure positions among the world's top 50 CO2 emitters. Iran ranks 18th, Saudi Arabia 22nd, and the UAE 43rd. Detail for all GCC countries with CO2 emissions is given below in table 1 (Global Carbon Project, 2021):

Table 1. Per capita CO2 emissions worldwide in 2020, by country (in metric tons)

S. No.	Country Name	Metric Tons
1	Bahrain	20.55
2	Kuwait	20.83
3	Oman	12.17
4	Qatar	37.02
5	Saudi Arabia	17.97
6	United Arab Emirates	15.19

GCC countries coordinate activities and set the framework for regional efforts on environmental issues through a number of regional and international organizations. Despite their relatively recent creation, environmental ministries, agencies, and councils are having difficulty playing a significant role in the decision-making process when it comes to addressing the GCC countries and the region's most pressing environmental issues, problems, and threats (Raouf, M., 2008). Table 2 shows the various institutions and agencies of GCC countries (Raouf, M., 2008):

Table 2. Governmental environmental institutions and agencies in GCC

Countries	Policy Institution	Executive Agency
Bahrain	Environment and Wildlife Affairs	Public Commission for the Protection of Marine Recourses, Environment, and Wildlife
Kuwait	Environment Public Authority	Environment Public Authority
Oman	Council of Ministers	Ministry Environment and Climate Change
Qatar	Council of Ministers (Permanent Commission for Environmental Protection)	Supreme Council for the Environment and Natural Resources
Saudi Arabia	Ministerial Committee on Environment	Presidency of Meteorology and Environment (PME)
United Arab Emirate	Council of federation	Federal Environmental Agency/Ministry of Environment and Water Resources

Source: Compiled from Global Carbon Project (Raouf, M., 2008)

In 2021, Saudi Arabia's strategy to diversify its economy through clean and renewable power assets was further strengthened with the financing of projects in the solar sector and the establishment of the National Infrastructure Fund (NIF). A substantial increase has been made by Saudi Arabia to the solar photovoltaics (PV) targets for 2023 and 2030, respectively, with a goal of 40 GW and 20GW. In April 2021, the kingdom announced the signing of power purchase agreements (PPAs) for approximately 3 GW (2,970 MW) of solar projects under the National Renewable Energy Program. As of mid-2021, Sudair, Jeddah, and Rabigh, together accounting for 2.1 GW, had reached financial close; both Jeddah and Rabigh are scheduled to come on line in 2022, and Sudair's first phase is scheduled to begin operating in 2023. With this additional capacity installed, the total solar capacity installed by 2023 will be about 2.5 GW or 12.5% of what was originally intended (Kiyasseh, L., 2022).

A target of 50% clean energy generation has been set by Saudi Arabia by 2030. As part of this initiative, both the ministries of energy and industry and minerals will pursue complementary tracks. The Renewable Energy Project Development Office (REPDO) will manage the procurement of 30% of this target through a competitive process; the kingdom's Public Investment Fund (PIF) will deliver the remaining 70% through direct consultations with investors. In spite of that, the kingdom took strong climate action in October, announcing a net-zero target for 2060 and updating its Nationally Determined Contributions under the Paris Agreement. To achieve its "Carbon Circular Economy" goals, Saudi Arabia pledged a total of $340 billion in net-zero investments to go to renewable energy, storage, and hydrogen projects - including carbon capture, utilization, and storage - that will help facilitate the country's transition to carbon-neutrality.

Since many GCC countries lack institutional capacity and have relatively new environmental authorities, they join international agreements first and then find ways to fulfill their commitments. Additionally, there have been numerous local developments, such as:

- In November 2007, Oman has renamed the Ministry of Environment and Regional Municipalities to the Ministry of Environment and Climate Change to reflect the importance of climate change.
- A Global Footprint Network initiative was launched by the Environment Authority in Abu Dhabi (AGEDI) in collaboration with WWF-UAE to collect and prepare the UAE's footprint in response to the lack of reliable data on environmental affairs.
- On October 18, 2007, the UAE launched Al Basama Al Beeiya (Ecological Footprint) Initiative to reduce its ecological footprint and sustainable development.
- During the first announcement of Saudi Arabia's Vision 2030 economic development plan, solar power was discussed a great deal as a means of curbing emissions and distancing electricity production from liquid fuel, in an effort to monetize its crude (Kiyasseh, L., 2022).

4. SOURCES OF RENEWABLE ENERGY

Renewable energy sources are those sources that are replacing fossil energy naturally. Technologies are progressing to generate electricity from a few common resources of renewable energy can be given as:

- **Wind Energy**

"Wind power" or "wind energy" are terms used to describe wind-generated electrical or mechanical power. Wind energy is a clean, renewable source of energy, produced by wind turbines. Wind turbines turn their blades when the wind blows against them. During wind turbine operation, blades turn around a rotor, which accelerates the blade rotation, which allows for a smaller generator to be used. The shaft and gears that speed up the blade rotation allow for the smaller generator. This process creates electricity. Table 3, given below shows installed wind power capacity for top countries (Renewable energy capacity worldwide by country, 2020):

Table 3. Installed wind power capacity for top 10 countries (MW)

S. No.	Country	2021
1	China	328,973
2	United States	132,738
3	Germany	63,760
4	India	40,067
5	Spain	27,497
6	United Kingdom	27,130
7	Brazil	21,161
8	France	18,676
9	Canada	14,304
10	Sweden	12,080

- **Solar Energy**

An energy source derived from the sun is solar power, which can be used for thermal or electrical purposes. Almost every country in the world has access to solar power, which is clean, cheap, and renewable. Sunlight is an endless source of energy since it comes directly from the sun. Table 4, shows installed solar power capacity for top countries (Renewable energy capacity worldwide by country, 2020):

Table 4. Installed solar power capacity for top 10 countries (MW)

S. No.	Country	MW
1	China	306,972
2	United States	95,208
3	Japan	74,191
4	Germany	58,461
5	India	49,684
6	Italy	22,698
7	Australia	19,076
8	Rep of Korea	18,160
9	Vietnam	16,660
10	Spain	15,952

- **Geothermal Energy**

Geothermal energy comes from the decay of uranium, thorium, and potassium radioactive isotopes in the natural heat of the earth. The Earth's surface heat flow averages 82 mW/m2, which amounts to 42 million megawatts of total heat. There is an enormous amount of thermal energy on the Earth, but only a fraction of it can be harnessed. The utilization of geothermal energy is currently limited to areas where geological conditions make it possible for a carrier (liquid or vapor phase water) to transfer heat from deep hot zones to or near the surface (World Energy Resources: Geothermal, 2013). Geothermal power capacity for top countries are (Renewable energy capacity worldwide by country, 2020):

- **Ocean Energy**

Tides, waves, and currents are the major sources of electricity from ocean energy. Tides are responsible to produce electricity using high and low tides. Wave energy, converters the energy from ocean waves and use it to generate electricity. An ocean thermal energy conversion unit produces power by converting the difference in temperature between warm surface seawater and cold seawater 800–1,000 meters below the surface (IRENA). Marine capacity for countries (Renewable energy capacity worldwide by country, 2020) can be given in Table 6 as:

Table 5. Installed geothermal power capacity for top 10 countries (MW)

S. No.	Country	MW
1	United States	3,889
2	Indonesia	2,276
3	Philippines	1,928
4	Turkey	1,676
5	New Zealand	984
6	Mexico	976
7	Kenya	863
8	Italy	802
9	Iceland	756
10	Japan	481

Table 6. Marine capacity worldwide for top countries (MW)

S. No.	Country	MW
1	South Korea	256
2	France	214
3	United Kingdom	22
4	Canada	20
5	China	5
6	Spain	5
7	Netherlands	2
8	Russia	2
9	Australia	1

- **Hydropower Energy**

It is the oldest and largest source of renewable energy. Hydropower uses the flow of water to produce energy. It is a clean source of energy. Hydropower energy is normally generated through a dam where water is flowing from a height towards a lower level. It is an inexpensive source of energy. Below is the table showing installed hydropower capacity for top countries (Renewable energy capacity worldwide by country, 2020):

Table 7. Installed hydropower capacity for top ten countries (MW)

S. No.	Country	MW
1	China	354,530
2	Brazil	109,426
3	Canada	82,562
4	United States	79,982
5	Russia	51,145
6	India	46,779
7	Norway	34,813
8	Turkey	32,492
9	Japan	28,125
10	France	23,985

- **Bio Energy**

Bioenergy is one of the miscellaneous sources of energy that meet our requirement for energy. It is the type of renewable energy which comes from the living materials known as biomass. It is used to produce heat, energy, fuels, and products. Table 8, shows installed bioenergy capacity (Renewable energy capacity worldwide by country, 2020):

Table 8. Installed bioenergy capacity for top ten countries (MW)

S. No.	Country	MW
1	China	29,753
2	Brazil	16,300
3	United States	13,573
4	India	10,591
5	Germany	10,449
6	United Kingdom	7,259
7	Japan	4,591
8	Sweden	4,461
9	Thailand	4,222
10	Italy	3,439

Renewable energy production increased from 2020 – to 2021. It can be seen in the chart below (IEA):

Figure 2. Renewable energy production by technology, country, and region, 2020 – 2021

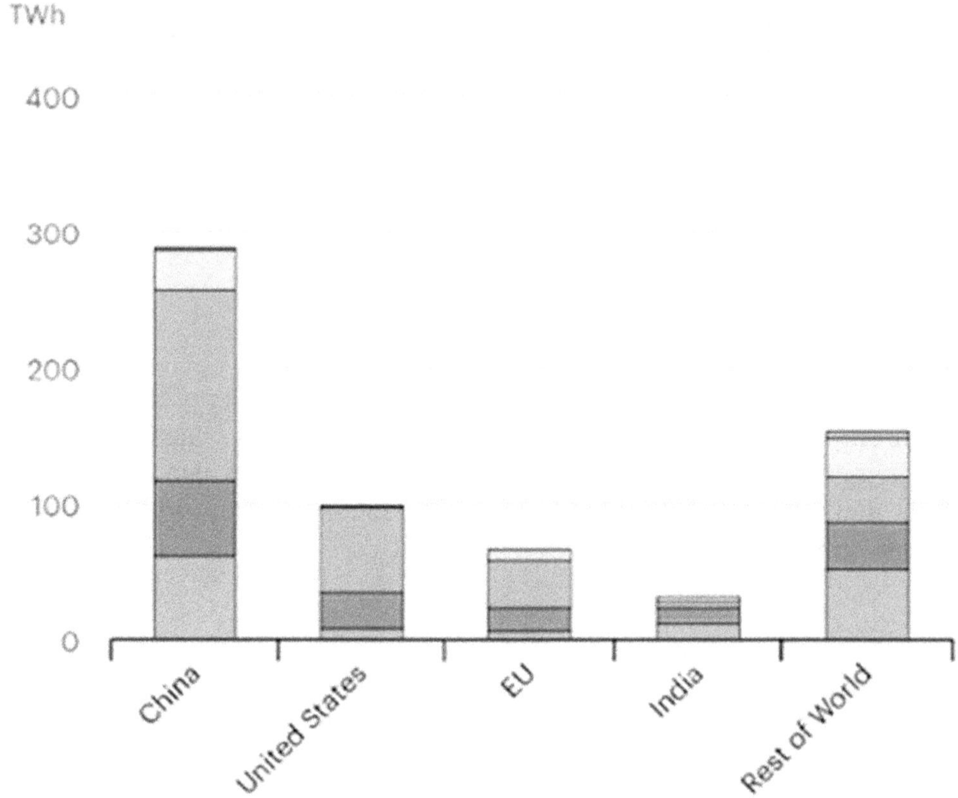

From the figure 2, we can explain that China is on top of the chart with production of hydro 62, solar 54, wind 141, bioenergy 30, and others 1 Tera watt-hour (Twh), The United States secured second place by producing hydro 9, solar 26, wind 61, and bioenergy 2 Twh, Europian Union (EU) takes the third position by producing hydro 7, solar 17, wind 35, and bioenergy 8 Twh, India gained secured the fourth position by producing hydro 12, solar 12, wind 5, and bioenergy 3 Twh, and Rest of world hydro 51, solar 36, wind 33, bioenergy 29, and others 1 Twh.

Below, in figure 3, is the comparison of renewable electricity generation increase by technology, 2019-2020 and 2020-2021 (IEA):

Figure 3. Renewable electricity generation increase by technology, 2019-2020 and 2020-2021

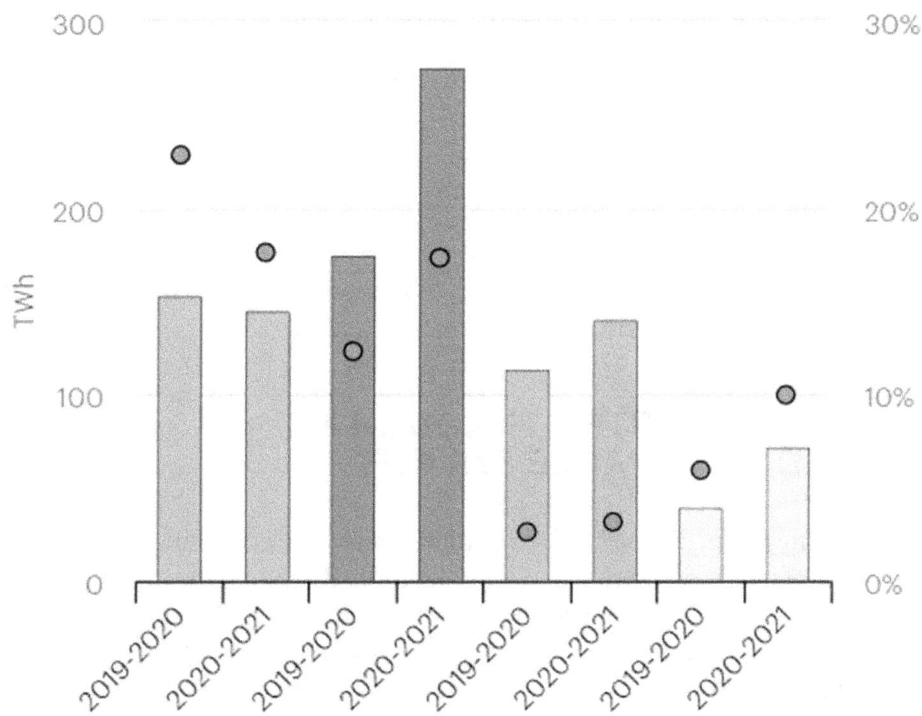

In the coming years, wind energy will increase by almost 17%, or 275 TWh, significantly more than in 2020. In 2021, China generated 600 TWh and the United States 400 TWh, together representing more than half of global wind energy production. Even with China retaining the largest PV market, the expansion in the United States will be supported by federal and state policy. The production of solar PV electricity grew by 145 TWh in 2021, almost 18%, to 1000 TWh. Thanks to incentives, waste-to-power projects in Asia will drive bioenergy growth (IEA).

In Table 9, we can see a Fast Fact – Climate and temperature (On climate and temperature rise). It shows the climate actions on climate and temperature rise:

Table 9. Fast fact: Climate and temperature

S. No.	Climate Actions
1.	2015-2019 was the warmest years on evidence while the warmest decade on documentation was 2010-2019.
2.	Over the last 2000 years, the global surface temperature has risen more rapidly since 1970 than during any other 50-year period.
3.	By the end of the century, the temperature could rise by 4.4°C based on current carbon dioxide emissions.
4.	CO_2 levels reached 148 percent above preindustrial levels in 2019 as greenhouse gas concentrations reached new highs.
5.	There has been a continued rise in greenhouse gas concentrations despite already being at their highest level in 2 million years.
6.	For temperatures to be kept below 2°C by 2030, emissions must decrease 7.6 percent per year from 2020 to 2030.
7.	Global warming is estimated to be below 2°C by 2030 if greenhouse gas emissions are reduced 15 gigatons carbon dioxide equivalent (Gt CO_2e) from current levels.
8.	By 2030, the world will have to reduce fossil fuel production by about 6 percent per year to follow a 1.5°C path.

5. GETTING SMARTER THROUGH ICT ENABLED RENEWABLE ENERGY

The increasing importance of ICT contributes to several economic sectors in the modern age of industrialization and globalization; therefore, the environmental performance of ICT cannot be ignored (Cheng et al., 2021). Economic literature cannot ignore the important role that ICT plays in sustainable development. In accordance with international organizations such as the United Nations, the International Telecommunications Union, the Organization for Economic Co-Operation and Development (OECD), and the World Bank, the ICT industry is an integral part of sustainable development from an integrated perspective (Toader et al., 2018). A wide range of industries consume and produce energy, including aerospace, construction, transportation, farming, mining, and the information and communication technology industry. An 'Earth temperature boost' and environmental change have received greater attention in recent years as concerns over global warming rise (Azam A. et al., 2021). Through the use of Distributed Renewable Energy Sources (DRESs), which complement or even replace centralized fossil fuels, the Smart Grid represents an important evolution of the electricity grid by shifting energy generation and consumption from a top-heavy, strictly hierarchical, and downstream system to a distributed bi-directional system. A distributed ICT infrastructure has been described by Stocker et al (2022) in order to overcome such challenges in the future, specifically related to Ancillary Services in Smart Grids. A Smart Grid Architecture Model (SGAM) framework has been used to develop the proposed infrastructure, as described in Smart Grid Mandate M/490 by the European Commission Stocker et al (2022).

The demand for energy is rising daily due to the increasing population and economic development. As per the increasing energy productivity and decreasing energy intensity of the European Union (EU) economy, Information and Communication Technologies (ICT) have been recognized to play a significant role. The use of ICT for energy efficiency will not only improve the efficiency of European industry but will also create a large market for ICT-enabled energy-efficiency technologies and services that will accelerate the European industry's competitiveness (World Energy Council, 2018).

The total decline in greenhouse gas required by 2020 was approximately from 50 percent to 125 percent. Approximately 40 percent of the final energy is consumed by the construction area which is the largest single energy utilizer in the EU. ICT products and services should be measured and analyzed using standardized methods. ICT solutions for energy-efficient management are monitored and evaluation should be done on a regular basis (World Energy Council, 2018).

Energy goals set for the coming decades by the EU are given in Figure 4 below:

Figure 4. Climate and energy framework of EU for 2030

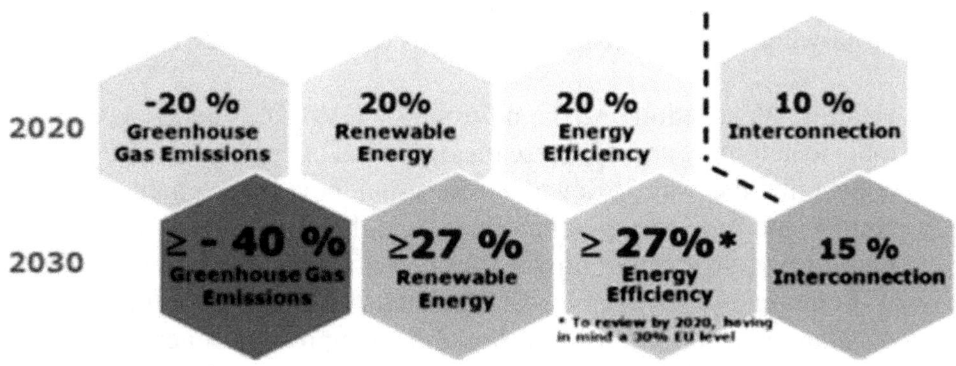

In 2012, the EU adopted the Energy Efficiency Directive, which establishes a set of binding measures that will help the EU reach its 20% energy efficiency target by 2020 (World Energy Council, 2018). As part of the proposed Clean Energy for all Europeans proposal, the European Commission updated the Energy Efficiency Directive, proposing a new 30% energy efficiency target for 2030, along with measures to ensure it is met (World Energy Council, 2018).

5.1 Energy Efficiency Improved by ICT

By reducing energy requirements for product or service delivery, ICT can contribute to an increase in energy efficiency. Energy efficiency can be improved in major energy-consuming sectors through the direct monitoring and management of energy consumption. By providing innovative technologies that optimize energy use according to actual demand, e.g., occupancy levels of buildings. ICT can decrease inefficient consumption of energy one perfect example is Solid-state lighting. In today's computer-based systems, there are many redundancies that can be reduced through emerging technologies such as thin client (computers without hard disk), grid computing, and virtualization (World Energy Council, 2018).

ICT can significantly assist in providing new opportunities such as ICT-based Neighborhood Management Systems that allow peer-to-peer sharing of power generated through renewable systems.

ICT can also look into the complications of measuring energy performance at a system level. There are software tools that can provide information about how to optimize its overall performance in a cost-effective way.

The main reductions in energy consumption by using ICT can be achieved in the following energy-consuming systems (Stallo et. al, 2010):

- Heating, Ventilation, Air Conditioning (HVAC) systems: Using ICT tools to automate and optimize the operation of space and water heating is an effective way to leverage the efficiency gains enabled by ICT control and monitoring capabilities. e.g., heating control, temperature monitoring, switchable mirror film on windows, integrated cooling of ICT equipment, switchable vacuum insulated panels, integrated control of clean room conditions). It is possible to improve the energy efficiency of buildings with ICT with a more controlled thermal performance;
- Lighting systems: A reduction in energy consumption of lighting systems is possible due to the use of ICT-based lighting technologies (e.g., LED lighting) and control systems (e.g., daylight sensors).

According to Christian Bruch, Siemens Energy Chief Executive Officer, and Chief Sustainability Officer, we have a massive and complex transformation to undergo, and I'll say it without any hesitancy. There is no doubt that the elements we are driving today are insufficient. As per the report of The IPCC (Intergovernmental Panel on Climate Change), We must move fast and in a different way. The time has come to shift gears and bring all stakeholders together (Sachgau, O., 2022).

5.2 Policies to Promote Energy ICT

The governments should take policy measures to increase and promote the share of renewable energy. There should be a national policy that includes a provision for renewable purchase obligations. Government should focus more and more on ICT-enabled solar and wind energy systems. It will be easier for governments to ensure a policy environment that is enabling with all relevant stakeholders consulted. Some of the policies should involve:

- Developments of solar parks and solar projects should be done.
- Focus should be to develop the green corridor and energy transmission should take place from the corridor.
- Policy should be made to support the rooftop solar plant for housing societies and isolated houses.
- Allowing maximum Foreign Direct Investment in the segment without any hurdle.
- Providing support for renewable energy research and development with industry participation.
- A policy should be aimed at accelerating the adoption of ICT-enabled renewable energy by addressing various green energy concerns.

In order for ICT experts to contribute to the development of society and its prospects, policymakers need to enable them to be trained by providing education, health services that broaden people's horizons, and competences that enable them to contribute to institutions that provide social facilities and economic incentives (Azam A. et al., 2021).

5.3 Justness of the Energy Transition

Currently, we are going through an energy transition, and we must ensure it is just and equitable. There is no inherent justice or unjustness in renewable energy technologies. A move to clean energy systems must also take equity and justice into account in addition to the implications of moving away from fossil fuels. Decarbonization, digitalization, and electrification all play a role in the energy transition. Transitions that are just and equitable reduce and do not exacerbate inequalities by distributing benefits and costs in an equitable manner. In order for a just transition to take place, human rights throughout the energy supply chain must be respected. Various stakeholders like government, civil society and other investors have to play an important role (Cabré & Araujo, 2022).

The concept of just and equitable energy transition may have many definitions and conceptual approaches, but we understand it as being one in which both the process and the end state effectively address justice and equity concerns as it transitions from fossil fuel-dominated energy systems to ones dominated by renewable energy sources (Cabré & Araujo, 2022).

The public's views on justice and fairness can delay or stall the development of renewable energy projects and clean energy supply chains. As seen in cases from the industry in both developed and developing countries (European Western Balkans, 2021; Segreto et al., 2020; Stigka et al., 2014), public support for renewable energy projects and those related to clean energy supply chains is crucial. Just and equitable energy transitions are important, but by themselves, a just transition cannot resolve all the world's injustices and inequities (Cabré & Araujo, 2022).

6. CHALLENGES AND CONSTRAINTS

In recent years renewable energy has attracted people more compared to the previous decades. Slowly, it is becoming the main source of energy. It is well known that ICT faces a number of obstacles and constraints before it can reach its full potential. Nevertheless, emerging factors will be more problematic in the future due to changes in commercial business models and underlying technologies. In the household sector, ICT plays an important role in promoting energy efficiency and reducing GHG emissions. Some of the challenges can be summarized as:

- High initial installation cost is a big challenge for undeveloped and developing countries (Moradiya, M., 2019).
- Comprehensive understanding of energy utilization and usage.
- Wind and solar generation can pose challenges to grid operators due to their uncertainty and variability. It is possible that wind generation will increase as load levels increase, but additional measures must be

taken if renewable generation increases when load levels decrease (or vice versa) (Bird et. al, 2013).

- It is more common for solar generation to coincide with a load than with wind, but when solar generation is lost at sunset, ramping requirements to meet evening demands may be exacerbated (Bird et. al, 2013).
- Climate, weather, and geography affect the availability of renewable energy sources, therefore, not every source is suitable for every region (Moradiya, M., 2019).
- The use of ICT and smart meters has facilitated greater visibility than ever before. However, there are still challenges with utilizing these advancements to their full potential.
- Energy efficiency will be impacted by the deployment of existing technology. In the UK, for instance, many households provide manual readings of their smart electric meters every month, whereas smart metering allows for real-time usage monitoring (World Energy Council, 2018).
- In India, financing renewable energy presents a unique set of challenges based largely on the structure and investment character of the industry, which in turn are impacted by the types of instruments and sentiments of investors.
- It continues to be a major challenge for the sector to attract the type and size of investment necessary due to its capital-intensive nature and private-sector-led development (Sarangi, G. K., 2018).
- With the growth of home automation and the improvement of the Internet of Things, home appliances will be able to provide a vast amount of data to enhance energy efficiency.
- Power quality is the major issue. Consistent higher power is required to ensure efficient working.
- Cybersecurity is also a big concern in renewable energy. Sharing and hacking of data is a major issue.

Due to the poverty and lack of access to project financing in developing countries, these countries may encounter the following constraints in implementing ICT solutions to mitigate climate change:

- As many developing countries are resource-constrained, the adoption of smart technologies is likely to be lax, which exacerbates GHG emissions.
- Considering that climate change mitigation applications are relatively less known, adoption of ICT is expected to be rather slow.

- Considering environmental conservation and climate change as irrelevant to economic development, the conservative banking sector support is insufficient.
- Many countries or organizations are unable to use technology that is compatible with local conditions, due to the limited or uncertain suitability of ICTs.

7. CONCLUSION AND RECOMMENDATIONS

This chapter illustrates how ICT applications can be used to reduce greenhouse gas emissions by reducing carbon dioxide emissions into the atmosphere, thus contributing to securing the future. Building climate resilience and adapting to climate change is becoming an increasingly important issue for countries worldwide. Energy conservation may not cause by the technology itself, but a rebound effect can cause energy savings, regardless of how the technology is deployed and used. In order to mitigate the effects of climate change caused by anthropogenic activity, energy and information technologies policies will be crucial. It is also possible to develop technology transfer and financing schemes that incorporate the broader deployment of ICTs in developing countries into carbon offset projects covered by The United Nations Framework Convention on Climate Change (UNFCCC) (Niyibizi, A., & Komakech, A., 2013). A number of environmental challenges face the GCC countries and they will have to juggle many competing priorities, from economic diversification, water supply, food security, environmental protection, and conservation to newly hazardous effects. Climate change continues to impact all components of the energy system, weather events impact the availability of primary energy sources, especially renewable sources as well as the transformation, transmission, distribution, and storage of energy. Several policy implications are recommended. Furthermore, there is a high possibility for added development using low-cost ICT applications, supported by large corporates.

In the midst of the energy transition, it is our responsibility and opportunity to ensure that, in addition to helping to create a healthier planet by replacing fossil fuels with clean energy sources, this is done in an equitable and just manner, which will lead to prosperity for everyone (Cabré & Araujo, 2022).

Recommendations

- ICT-enabled carbon offset projects must be designed and developed in less developed countries in order to maximize the benefits of emissions reductions while also promoting sustainable socio-economic development, especially in rural areas.
- For ICT-enabled carbon offset projects to succeed, it is imperative that local commercial banks and microfinance institutions (MFIs) expand their lending capabilities.
- Usually, these projects are centered on energy efficiency and renewable energy as substitutes for solid energy options such as biomass; the process may be complex, as it also requires the banking sector to be trained to evaluate the social and economic benefits of carbon offsets.
- A reduction in GHG emissions, the control of heat and floods, as well as incentives for offsetting ambitious targets, are all necessary.
- Academics and researchers must lead the way in finding novel technological solutions that can mitigate climate change, while businesses and non-government actors (the private sector) can incubate these solutions into tangible products, while governments must facilitate technology incubation for the private sector.
- The industrial, energy, construction, and transportation sectors in developing nations need to embrace low-carbon development. Enhanced innovation capacity in ICT-enabled carbon offset projects will also contribute to achieving this goal.
- Development projects that include carbon offset measures should be prioritized for immediate funding by the national planning authorities. A regulatory, verification and certification standard should be developed by policymakers, or existing ones should be modified based on industry best practices.
- A strong policy advocacy effort is needed to support regulatory and policy reforms for better investment opportunities in ICT-enabled carbon offset projects. Building the capacity of non-government actors to oversee compliance with climate-proofed development ICTs will be beneficial.

REFERENCES

Abu, T., & Bressler, L. (2019). Energy security in Israel and Jordan: The role of renewable energy sources. *Renewable Energy*, *135*, 378–389. doi:10.1016/j.renene.2018.12.036

Ahmed, F., Naeem, M., & Iqbal, M. (2016). *ICT and renewable energy: a way forward to the next generation telecom base stations*. Telecommun Syst. doi:10.100711235-016-0156-4

Azam, A., Rafiq, M., Shafique, M., Yuan, J., & Salem, S. (2021). Human Development Index, ICT, and Renewable Energy-Growth Nexus for Sustainable Development: A Novel PVAR Analysis. *Frontiers in Energy Research*, *9*, 760758. doi:10.3389/fenrg.2021.760758

Baburajan, R. (2022). *MWC 2022: ICT is not a contributor to carbon footprint, but an enabler of carbon handprints, says Huawei.* TelecomLead. https://www.telecomlead.com/telecom-equipment/mwc-2022-ict-is-not-a-contributor-to-carbon-footprint-but-an-enabler-of-carbon-handprints-says-huawei-103683

Benighaus, C. (2019). Neither risky technology nor renewable electricity: Contested frames in the development of geothermal energy in Germany. *Energy Research & Social Science, 47*(August 2018), 46–55.

Bird, L., Milligan, M., & Lew, D. (2013). *Integrating Variable Renewable Energy: Challenges and Solutions. National Renewable Energy Laboratory, Technical Report*. NREL. https://www.nrel.gov/docs/fy13osti/60451.pdf

Borowski, P. F. (2021). Innovative Processes in Managing an Enterprise from the Energy and Food Sector in the Era of Industry 4.0. *Processes (Basel, Switzerland)*, *9*(2), 381. doi:10.3390/pr9020381

BP energy outlook 2035. (n.d.). *BP Energy Outlook.* India Environment Portal. http://www.indiaenvironmentportal.org.in/content/406084/bp-energy-outlook-2035/

Bronk, C., Lingamneni, A., & Palem, K. (2010). *Innovation for sustainability in information and communication technologies (ICT).* Internal report, Rice University. https://www.bakerinstitute.org/media/files/Research/dbfd5eba/ITP-pub-SustainabilityInICT-102510.pdf

Cabré, M. M., & Araujo, J. V. (2022). *Considerations for a just and Equitable Energy Transition*. SEI (Stockholm Environment Institute). https://www.sei.org/publications/just-equitable-energy-transition/ (Accessed: 31 July 2023).

Céspedes-lorente, J. J., & Magán-díaz, A. (2016, April). (2018). Information technologies and downsizing: Examining their impact on economic performance. *Information & Management*. doi:10.1016/j.im.2018.09.012

Cheng, C., Ren, X., Dong, K., Dong, X., & Wang, Z. (2021). How Does Technological Innovation Mitigate CO2 Emissions in OECD Countries? Heterogeneous Analysis Using Panel Quantile Regression. *Journal of Environmental Management*, *280*, 111818. doi:10.1016/j.jenvman.2020.111818 PMID:33360390

Coleman, S. (2005). New Mediation and Direct Representation: Reconceptualizing Representation in the Digital Age. *New Media & Society*, *7*(2), 177–198. doi:10.1177/1461444805050745

Erumban, A. A., & Kusum, D. (2016). Information and communication technology and economic growth in India. *Telecommunications Policy*, *40*(5), 412–431. doi:10.1016/j.telpol.2015.08.006

European Western Balkans. (2021, November 28). *Environmental protests and roadblocks across Serbia, masked men attack citizens*. European Western Balkans. https://europeanwesternbalkans.com/2021/11/28/environmental-protests-androadblocks-across-serbia-masked-men-attack-citizens/

García-Quismondo, E., Santos, C., Lado, J., Palma, J., & Anderson, M. A. (2013). Optimizing the Energy Efficiency of Capacitive Deionization Reactors Working under Real-World Conditions. *Environmental Science & Technology*, *47*(20), 11866–11872. doi:10.1021/es4021603 PMID:24015835

Global Carbon Project. (2021). *Per capita CO2 emissions worldwide in 2020, by country (in metric tons)* [Graph]. In Statista. https://www.statista.com/statistics/270508/co2-emissions-per-capita-by-country/

Glossary. (2001). *Climate Change. Education Center – Arctic Climatology and Meteorology*. NSIDC National Snow and Ice Data Center.

Hache, E. (2018). Do renewable energies improve energy security in the long run? *Inter Economics*, *156*, 127–135. doi:10.1016/j.inteco.2018.01.005

IEA. (2021). *Renewable electricity generation increase by technology, country and region*. IEA, Paris. https://www.iea.org/data-and-statistics/charts/renewable-electricity-generation-increase-by-technology-country-and-region-2020-2021

ITU. (2015). Focus Group on Smart Sustainable Cities. *Information and communication technologies for climate change adaptation in cities*. UNFCCC. https://www4.unfccc.int/sites/NAPC/Documents/Supplements/ICTs-for-climate-change-adaptation.pdf

Jorgenson, D. W., Ho, M. S., & Samuels, J. D. (2016). The impact of information technology on postwar US economic growth. *Telecommunications Policy*, *40*(5), 398–411. doi:10.1016/j.telpol.2015.03.001

Kiyasseh, L. (2022). *Strong momentum in Saudi Arabia's drive toward renewables and infrastructure*. Middle East Institute. https://www.mei.edu/publications/strong-momentum-saudi-arabias-drive-toward-renewables-and-infrastructure

Lin, B., & Chen, Y. (2019). Does electricity price matter for innovation in renewable energy technologies in China? *Energy Economics*, *78*, 259–266. doi:10.1016/j.eneco.2018.11.014

Migliaccio, G. (2016). ICT for Disability Management in the Net Economy. *International Journal of Globalisation and Small Business*, *8*(1), 51. doi:10.1504/IJGSB.2016.076452

Moradiya, M. (2019). *The Challenges Renewable Energy Sources Face*. AZoCleantech. https://www.azocleantech.com/article.aspx?ArticleID=836

Niyibizi, A., & Komakech, A. (2013). Climate Change Mitigation in Developing Countries Using ICT as an Enabling Tool. In W. Leal Filho, F. Mannke, R. Mohee, V. Schulte, & D. Surroop (Eds.), *Climate-Smart Technologies. Climate Change Management*. Springer. doi:10.1007/978-3-642-37753-2_2

Oliveira, T. A., Oliver, M., & Ramalhinho, H. (2020). Challenges for Connecting Citizens and Smart Cities: ICT, E-Governance and Blockchain. *Sustainability (Basel)*, *12*(7), 2926. doi:10.3390u12072926

Palvia, P., Baqir, N., & Nemati, H. (2015). ICT Policies in Developing Countries: AN Evaluation with the Extended Design-Actuality Gaps Framework. *The Electronic Journal on Information Systems in Developing Countries*, *71*(1), 1–34. doi:10.1002/j.1681-4835.2015.tb00510.x

Pfeifer, A., Dobravec, V., Pavlinek, L., & Kraja, G. (2018). Integration of renewable energy and demand response technologies in interconnected energy systems. *Energy*, *161*, 447–455. doi:10.1016/j.energy.2018.07.134

Pradhan, R. P., Arvin, M. B., & Norman, N. R. (2015). The dynamics of information and communications technologies infrastructure, economic growth, and financial development: Evidence from Asian countries. *Technology in Society*, *42*, 135–149. doi:10.1016/j.techsoc.2015.04.002

Pradhan, R. P., Mallik, G., & Bagchi, T. P. (2018). Information communication technology (ICT) infrastructure and economic growth: A causality evinced by cross-country panel data. *IIMB Management Review*, *30*(1), 91–103. doi:10.1016/j.iimb.2018.01.001

Rahman, M., & Mezbah-ul-Islam, M. (2012). *Issues and Challenges for Sustainable Digital Preservation Practices in Bangladesh*. Research Gate. https://www.researchgate.net/publication/261178681_Issues_and_Challenges_for_Sustainable_Digital_Preservation_Practices_in_Bangladesh

Raouf, M. (2008). *Climate Change Threats, Opportunities, and the GCC Countries*. Middle East Institute. https://www.mei.edu/publications/climate-change-threats-opportunities-and-gcc-countries

Renewable energy capacity worldwide by country. (2020). Statista. https://www.statista.com/statistics/267233/renewable-energy-capacity-worldwide-by-country/

Renewable energy – powering a safer future. (2022). United Nations. https://www.un.org/en/climatechange/raising-ambition/renewable-energy

Ribeiro, F., Ferreira, P., Araújo, M., & Braga, A. C. (2018). Modeling perception and attitudes towards renewable energy technologies. *Renewable Energy*, *122*, 688–697. doi:10.1016/j.renene.2018.01.104

Riti, J. S., Song, D., Shu, Y., Kamah, M., & Atabani, A. A. (2018). Does Renewable Energy Ensure Environmental Quality in Favour of Economic Growth? Empirical Evidence from China's Renewable Development. *Quality & Quantity*, *52*(5), 2007–2030. doi:10.100711135-017-0577-5

Rot, A., Sobińska, M., Hernes, M., and Franczyk, B. (2020). Digital Transformation of Public Administration through Blockchain Technology. *Towards Industry 4.0—Current Challenges in Information Systems*. Springer. doi:10.1007/978-3-030-40417-8_7

Saberi, S., Kouhizadeh, M., Sarkis, J., & Shen, L. (2019). Blockchain Technology and its Relationships to Sustainable Supply Chain Management. *International Journal of Production Research*, *57*(7), 2117–2135. doi:10.1080/00207543.2018.1533261

Sachgau, O. (2022). *How to fight climate change now.* Siemens-Energy. https://www.siemens-energy.com/global/en/news/magazine/2022/how-to-fight-climate-change-now.html

Sarangi, G. K. (2018). *Green energy finance in India: Challenges and solutions.* Asian Development Bank. https://www.adb.org/publications/green-energy-finance-india-challenges-and-solutions

Segreto, M., Principe, L., Desormeaux, A., Torre, M., Tomassetti, L., Tratzi, P., Paolini, V., & Petracchini, F. (2020). Trends in social acceptance of renewable energy across Europe: A literature review. *International Journal of Environmental Research and Public Health, 17*(24), 9161. doi:10.3390/ijerph17249161 PMID:33302464

Stallo, C., De Sanctis, M., Ruggieri, M., Bisio, I., & Marchese, M. (2010) ICT Applications in Green and Renewable Energy Sector. *Workshops on Enabling Technologies: Infrastructure for Collaborative Enterprises.*

Stigka, E. K., Paravantis, J. A., & Mihalakakou, G. K. (2014). Social acceptance of renewable energy sources: A review of contingent valuation applications. *Renewable & Sustainable Energy Reviews, 32*, 100–106. doi:10.1016/j.rser.2013.12.026

Stocker, A., Alshawish, A., Bor, M., Vidler, J., Gouglidis, A., Scott, A., Marnerides, A., De Meer, H., & Hutchison, D. (2022). An ICT architecture for enabling ancillary services in Distributed Renewable Energy Sources based on the SGAM framework. *Energy Informatics, 5*(1), 5. doi:10.118642162-022-00189-5

The United Nations Framework Convention on Climate Change. (1994). *About.* UNFCCC. https://unfccc.int/process-and-meetings/the-convention/what-is-the-united-nations-framework-convention-on-climate-change

Thompson, M. (2018). Social capital, innovation, and economic growth. *Journal of Behavioral and Experimental Economics, 73*, 46–52. doi:10.1016/j.socec.2018.01.005

Toader, E., Firtescu, B., Roman, A., & Anton, S. (2018). Impact of Information and Communication Technology Infrastructure on Economic Growth: An Empirical Assessment for the EU Countries. *Sustainability (Basel), 10*(10), 3750. doi:10.3390u10103750

Wang, K., Yan, X., Yuan, Y., & Tang, D. (2018). Optimizing Ship Energy Efficiency: Application of Particle Swarm Optimization Algorithm. Proc. Institution Mech. Eng. Part M J. *Proceedings of the Institution of Mechanical Engineers. Proceedings Part M, Journal of Engineering for the Maritime Environment*, *232*(4), 379–391. doi:10.1177/1475090216638879

World Energy Council. (2018). *The Role of ICT in Energy Efficiency Management: Household Sector*. World Energy Council. https://www.worldenergy.org/assets/downloads/20180420_TF_paper_final.pdf

World Energy Resources. (2013). *Geothermal*. World Energy Council 2013. https://www.worldenergy.org/assets/images/imported/2013/10/WER_2013_9_Geothermal.pdf

Yang, Y., Ren, J., Stubbe, H., Xu, D. & Tien, T. (2018). Using multi-criteria analysis to prioritize renewable energy home heating technologies. CBS Research Portal.

Zacharoula, S. A. (2012). Green Informatics: ICT for Green and Sustainability. *Agrárinformatika Folyóirat*, *3*(2). http://real.mtak.hu/23913/1/89_371_1_PB_u.pdf

Zeeshan, M., Han, J., Rehman, A., Ullah, I., & Mubashir, M. (2022). Exploring the Role of Information Communication Technology and Renewable Energy in Environmental Quality of South-East Asian Emerging Economies. *Frontiers in Environmental Science*, *10*, 917468. doi:10.3389/fenvs.2022.917468

Glossary

Adaptation: Any activity that reduces the vulnerability of human or natural systems to the impacts of climate change and climate-related risks, by maintaining or increasing adaptive capacity and resilience.

Blockchain: A distributed and immutable ledger that stores data in blocks and are linked together via cryptography.

Circularity: A circular economy (also referred to as "circularity") is an economic system that tackles global challenges like climate change, biodiversity loss, waste, and pollution. Most linear economy businesses take a natural resource and turn it into a product that is ultimately destined to become waste because it has been designed and made. This process is often summarised by "take, make, waste." By contrast, a circular economy uses reuse, sharing, repair, refurbishment, remanufacturing, and recycling to create a closed-loop system, minimize resource inputs, and create waste, pollution, and carbon emissions. The circular economy aims to keep products, materials, equipment, and infrastructure in use for longer, thus improving the productivity of these resources. Waste materials and energy should become input for other processes through waste valorization: either as a component or recovered resource for another industrial process or as regenerative resources for nature (e.g., compost). This regenerative approach contrasts with the traditional linear economy, which has a "take, make, dispose of" production model.

Climate Change: Refers to the increase in global warming and long term shifts in weather due to human activities.

Decentralized Energy: Energy that is generated off the main grid.

Digital Technologies: All electronic tools and devices and automatic systems that generate store or process information or data. It includes the Internet of Things, Artificial intelligence, and blockchain.

Distributed Energy: Production of electricity by a large number of small generators which can be solar roofs or small wind turbines.

Eco Commerce: Eco commerce is a business, investment, and technology-development model that employs market-based solutions to balancing the world's energy needs and environmental integrity. Through green trading and green finance, eco-commerce promotes the further development of "clean technologies" such as wind power, solar power, biomass, and hydropower.

Eco-Tariffs: An Eco-tariff, also known as an environmental tariff, is a trade barrier erected to reduce pollution and improve the environment. These trade barriers may take the form of import or export taxes on products with a large carbon footprint or imported from countries with lax environmental regulations.

Emissions Trading: Emissions trading (also known as cap and trade, emissions trading scheme, or ETS) is a market-based approach to controlling pollution by providing economic incentives for reducing the emissions of pollutants.

Environmental Enterprise: An environmental enterprise is an environmentally friendly/compatible business. Specifically, an environmental enterprise is a business that produces value in the same manner which an ecosystem does, neither producing waste nor consuming unsustainable resources. In addition, an environmental enterprise rather finds alternative ways to produce one's products instead of taking advantage of animals for the sake of human profits. To be closer to being an environmentally friendly company, some environmental enterprises invest their money to develop or improve their technologies which are also environmentally friendly. In addition, environmental enterprises usually try to reduce global warming, so some companies use environmentally friendly materials to build their stores. They also set in environmentally friendly place regulations. All these efforts of the environmental enterprises can bring positive effects both for nature and people. The concept is rooted in the well-enumerated theories of natural capital, the eco-economy, and cradle-to-cradle design. Examples of environmental enterprises would be Seventh Generation, Inc., and Whole Foods.

Green Economy: A green economy is an economy that aims at reducing environmental risks and ecological scarcities and that aims for sustainable development without degrading the environment. It is closely related to ecological economics but has a more politically applied focus. The 2011 UNEP Green Economy Report argues "that to be green, and an economy must be not only efficient but also fair. Fairness implies recognizing global and country-level equity dimensions, particularly in assuring a Just Transition to an economy that is low-carbon, resource-efficient, and socially inclusive."

Green Politics: Green politics, or ecopolitics, is a political ideology that aims to foster an ecologically sustainable society often, but not always, rooted in environmentalism, nonviolence, social justice, and grassroots democracy. It began taking shape in the western world in the 1970s; since then, Green parties have developed and established themselves in many countries around the globe and have achieved some electoral success.

Low-Carbon Economy: A low-carbon economy (LCE) or decarbonized economy is based on low-carbon power sources with minimal greenhouse gas (GHG) emissions into the atmosphere, specifically carbon dioxide. GHG emissions due to anthropogenic (human) activity are the dominant cause of observed climate change since the mid-20th century. Continued emission of greenhouse gases may cause long-lasting changes worldwide, increasing the likelihood of severe, pervasive, and irreversible effects for people and ecosystems.

Mini-Grid: an electricity distribution network involving a small-scale electricity generation having a power rating of less than 15MW and disconnected from national electricity transmission networks.

Mitigation: Any Activity that reduces the possible long-term impacts of climate change on human life and the eco-system, through the reduction in emissions of greenhouse gasses.

Natural Resource Economics: Natural resource economics deals with the supply, demand, and allocation of the Earth's natural resources. One main objective of natural resource economics is to understand better the role of natural resources in the economy to develop more sustainable methods of managing those resources to ensure their future generations. Resource economists study interactions between economic and natural systems intending to develop a sustainable and efficient economy.

Peer-To-Peer Energy Trading: A business model, based on an interconnected platform, that serves as an online marketplace where consumers and producers meet to trade electricity directly, without the need for an intermediary.

Policy: A set of ideas or plan for action adopted and implemented by any government entities to govern and rule one sector or particular situation.

Regulation: A set of rules, laws, directives, requirements, and restrictions adopted and applied by a government authority in order to manage and control a specific sector, practice, or situation.

Sustainable Development: Sustainable development is an organizing principle for meeting human development goals while simultaneously sustaining the ability of natural systems to provide the natural resources and ecosystem services on which the economy and society depend. The desired result is a state of society where living conditions and resources are used to continue to meet human needs without undermining the integrity and stability of the natural system. Sustainable development can be defined as development that meets the needs of the present without compromising the ability of future generations to meet their own needs. Sustainability goals, such as the current UN-level Sustainable Development Goals, address the global challenges, including poverty, inequality, climate change, environmental degradation, peace, and justice.

Compilation of References

Abam, F. I., Nwankwojike, B. N., Ohunakin, O. S., & Ojomu, S. A. (2014). Energy resource structure and on-going sustainable development policy in Nigeria: A review. *International Journal of Energy and Environmental Engineering*, *5*(1), 91–102. doi:10.100740095-014-0102-8

Abu, T., & Bressler, L. (2019). Energy security in Israel and Jordan: The role of renewable energy sources. *Renewable Energy*, *135*, 378–389. doi:10.1016/j.renene.2018.12.036

Adamu, M. B., Adamu, H., Ade, S. M., & Akeh, G. I. (2016). Household Energy Consumption in Nigeria: A Review on the Applicability of the Energy Ladder Mode. *Journal of Applied Science & Environmental Management*, *24*(2), 237–244. doi:10.4314/jasem.v24i2.7

Adaramola, M. S., & Oyewola, M. O. (2011). Wind speed distribution and characteristics in Nigeria. *ARPN Journal of Engineering and Applied Science*, *6*, 82–86.

Adejonwo, O. (2018). Nigeria's commitments under the climate change Paris Agreement: legislative and regulatory imperatives towards ensuring sustainable development. *4th Scientific Conference of the Association of Environmental Law Lecturers from African Universities in cooperation with the Climate Policy and Energy Security Programme for Sub-Saharan Africa of the Konrad- Adenauer-Stiftung and UN Environment*. UN Environment.

Adesete, A. A., Olanubi, O. E., & Dauda, R. O. (2022). Climate change and food security in selected Sub-Saharan African Countries. *Environment, Development and Sustainability*. doi:10.100710668-022-02681-0 PMID:36186913

Adger, W. N. (2006). Vulnerability. *Global Environmental Change*, *16*(3), 268–28. doi:10.1016/j.gloenvcha.2006.02.006

Afaha, S. J., & Ifarajimi, G. D. (2021). Energy poverty, climate change and economic growth. *African Journal of Economic and Sustainable Development*, *4*, 98–115. doi:10.52589/AJESD-U3LCOY0P

AfDB. (2020). Developing Africa's Workforce for the Future. *Economic Outlook*, 2020.

African Union Energy Commission. (n.d.). *Sao Tome and Principe Energy Balance 2020*. AU AFREC. https://au-afrec.org/en/central-africa/sao-tome-and-principe

Afsah, S., Laplante, B., & Wheeler, D. (1996). Controlling Industrial Pollution: a new paradigm. *World Bank policy research working paper*, (1672). World Bank.

Ahl, A., Yarime, M., Tanaka, K., & Sagawa, D. (2019). Review of blockchain-based distributed energy: Implications for institutional development. In *Renewable and Sustainable Energy Reviews* (Vol. 107, pp. 200–211). Elsevier Ltd. doi:10.1016/j.rser.2019.03.002

Ahmed, F., Naeem, M., & Iqbal, M. (2016). *ICT and renewable energy: a way forward to the next generation telecom base stations*. Telecommun Syst. doi:10.100711235-016-0156-4

Akrofi, M. M. (2021b). Balancing Post-COVID-19 Economic Growth with Renewable Energy Development in Oil-producing African Economies: Nigeria in Perspective. *Journal of International Affairs*. https://jia.sipa.columbia.edu/online-articles/balancing-post-covid-19-economic-growth-renewable-energy-development-oil-producing

Akrofi, M. M. (2021a). An analysis of energy diversification and transition trends in Africa. *International Journal of Energy and Water Resources*, *5*(1), 1–12. doi:10.100742108-020-00101-5

ALER. (2020). *Renewable Energy and Energy Efficiency in São Tomé and Príncipe - National Status Report*. ALER. https://www.lerenovaveis.org/contents/lerpublication/aler-relatorio-stp-nov2020.pdf

Alladi, T., Chamola, V., Rodrigues, J. J. P. C., & Kozlov, S. A. (2019). Blockchain in smart grids: A review on different use cases. In Sensors (Switzerland) (Vol. 19, pp. 4862). MDPI AG. doi:10.339019224862

Amanchukwu, R. N., Amadi-Ali, T. G., & Ololube, N. P. (2015). Climate change education in Nigeria: The role of curriculum review. *Education*, *5*(3), 71–79.

Ampaire, E. L., Jassogne, L., Providence, H., Acosta, M., Twyman, J., Winowiecki, L., & Van Asten, P. (2017). Institutional challenges to climate change adaptation: A case study on policy action gaps in Uganda. *Environmental Science & Policy*, *75*, 81–90. doi:10.1016/j.envsci.2017.05.013

Anderson, T. W., & Hsiao, C. (1981). Estimation of Dynamic Models with Error Components. *Journal of the American Statistical Association*, *76*(375), 598–606. doi:10.1080/01621459.1981.10477691

Apergis, N. (2004). Inflation, output Growth, Volatility and Causality: Evidence from Panel Data and the G7 Countries. *Economics Letters*, *83*(2), 185–191. doi:10.1016/j.econlet.2003.11.006

Arellano, M., & Bond, S. (1991). Some Tests of Specification for Panel Data: Monte Carlo Evidence and an Application to Employment Equations, Review of Economic Studies. *Blackwell Publishing*, *58*(2), 277–297.

Armbrust, M., Fox, A., Griffith, R., Joseph, A. D., Katz, R., Konwinski, A., Lee, G., Patterson, D., Rabkin, A., Stoica, I., & Zaharia, M. (2010). A view of cloud computing. In Communications of the ACM (Vol. 53, pp. 50–58). doi:10.1145/1721654.1721672

Atukeren, E. (2007). A causal analysis of the R&D interactions between the EU and the US. Global. *Economic Journal (London)*, *7*(4), 1850121.

Awaworyi, C. S., Smyth, R., & Trong-Anh, T. (2022). *Energy poverty, temperature and climate change*. Research Gate. https://www.researchgate.net/publication/360001563_Energy_poverty_temperature_and_climate_change

Awosope, C. A. (2020). Nigerian Electricity Industry: Issues, Challenges and Solutions. *Public Lecture Series*, *3*(2), 2–14.

Ayers, J. (2009). International funding to support urban adaptation to climate change. *Environment and Urbanization*, *21*(1), 225–240. doi:10.1177/0956247809103021

Ayoade, M. (2018). *Bridging the gap between climate change and energy policy options: What next for Nigeria? 4th Scientific Conference of the Association of Environmental Law Lecturers from African Universities in cooperation with the Climate Policy and Energy Security Programme for Sub-Saharan Africa of the Konrad- Adenauer-Stiftung and UN Environment*. UN Environment.

Azam, A., Rafiq, M., Shafique, M., Yuan, J., & Salem, S. (2021). Human Development Index, ICT, and Renewable Energy-Growth Nexus for Sustainable Development: A Novel PVAR Analysis. *Frontiers in Energy Research*, *9*, 760758. doi:10.3389/fenrg.2021.760758

Azzuni, A., & Breyer, C. (2020). Global energy security index and its application on national level. *Energies*, *13*(10), 2502. doi:10.3390/en13102502

Baburajan, R. (2022). *MWC 2022: ICT is not a contributor to carbon footprint, but an enabler of carbon handprints, says Huawei.* TelecomLead. https://www.telecomlead.com/telecom-equipment/mwc-2022-ict-is-not-a-contributor-to-carbon-footprint-but-an-enabler-of-carbon-handprints-says-huawei-103683

Bakar, L. (2015). *Governing Electricity in South Africa: Wind, Coal and Power Struggles*. University of East Anglia.

Baltagi, B. H. (2021). *Econometric analysis of panel data*. Springer Nature. doi:10.1007/978-3-030-53953-5

Bankymoon. (2020). *Blockchain enabled solutions and services*. http://bankymoon.co.za/

Barr, R., Fankhauser, S., & Hamilton, K. (2010). Adaptation investments: A resource allocation framework. *Mitigation and Adaptation Strategies for Global Change*, *15*(8), 843–858. doi:10.100711027-010-9242-1

Benighaus, C. (2019). Neither risky technology nor renewable electricity: Contested frames in the development of geothermal energy in Germany. *Energy Research & Social Science, 47*(August 2018), 46–55.

Bhattarai, M., & Hammig, M. (2001). Institutions and the environmental Kuznets curve for deforestation: A crosscountry analysis for Latin America, Africa and Asia. *World Development*, *29*(6), 995–1010. doi:10.1016/S0305-750X(01)00019-5

Bijoy, C. R., Chakma, A., Guillao, J. A., Hien, B., Indrarto, G. B., Lim, T., Min, N. E., Rai, T. B., Smith, O. A., Rattanakrajangsri, K., Thuy, H. X., & Zunga, U. (2022). *Nationally Determined Contributions in Asai: Do governments recognise the rights, roles and contributions of Indigenous Peoples?* Asia Indigenous People Pact Report. https://aippnet.org/nationally-determined-contribution-asia-governments-recognizing-rights-roles-contributions-indigenous-peoples/

Bird, L., Milligan, M., & Lew, D. (2013). *Integrating Variable Renewable Energy: Challenges and Solutions. National Renewable Energy Laboratory, Technical Report*. NREL. https://www.nrel.gov/docs/fy13osti/60451.pdf

Blimpo, M. P., & Cosgrove-Davies, M. (2019). Electricity Access in Sub-Saharan Africa: Uptake, Reliability, and Complementary Factors for Economic Impact. World Bank & AFD.

BNamericas. (2020). Brazil power generation back to growth. *BN Americas*. https://www.bnamericas.com/en/news/brazil-power-generation-back-to-growth

Bobrow, D. B. (2006). Policy design: Ubiquitous, necessary and difficult. In *Handbook of Public Policy* (pp. 75–96). SAGE Publications Inc. doi:10.4135/9781848608054.n5

Booth, D. E. (1995). Economic democracy as an environmental measure. *Ecological Economics*, *12*(3), 225–236. doi:10.1016/0921-8009(94)00046-X PMID:12292356

Borowski, P. F. (2021). Innovative Processes in Managing an Enterprise from the Energy and Food Sector in the Era of Industry 4.0. *Processes (Basel, Switzerland)*, *9*(2), 381. doi:10.3390/pr9020381

Bouwer, L. M., & Aerts, J. C. (2006). Financing climate change adaptation. *Disasters*, *30*(1), 49–63. doi:10.1111/j.1467-9523.2006.00306.x PMID:16512861

Bowen, A., Cochrane, S., & Fankhauser, S. (2012). Climate change, adaptation and economic growth. *Climatic Change, 113*(2), 95–106. doi:10.100710584-011-0346-8

BP energy outlook 2035. (n.d.). *BP Energy Outlook.* India Environment Portal. http://www.indiaenvironmentportal.org.in/content/406084/bp-energy-outlook-2035/

Bradshaw, A. (1997). Sense and sensibility: debates and developments in socio-legal research methods. In P. A. Thomas (Ed.), *Socio-legal studies* (pp. 99–122). Dartmouth Publishing Company. https://link.springer.com/article/10.1007/s10991-010-9069-6

Bronk, C., Lingamneni, A., & Palem, K. (2010). *Innovation for sustainability in information and communication technologies (ICT).* Internal report, Rice University. https://www.bakerinstitute.org/media/files/Research/dbfd5eba/ITP-pub-SustainabilityInICT-102510.pdf

Brooks, N., Anderson, S., Burton, I., Fisher, S., Rai, N., & Tellam, I. (2013). *An operational framework for tracking adaptation and measuring development*. TAMD.

Brown, H. C. (2011). Gender, climate change and REDD+ in the Congo Basin forests of Central Africa. *International Forestry Review, 13*(2), 163–176. doi:10.1505/146554811797406651

Buckley, D. (2014). Mobilising Adaptation Finance: The status of public finance related to national funding for developing countries. Presented for the Second Forum of the Standing Committee on Finance 21 June 2014. UNFCCC. https://unfccc. int/sites/default/files/s3_1_daniel_b_scf_adaptation_finance_seminar_undp_dbuckley.pdf

Burmeister, H., Cochu, A., Hausotter, T., & Stahr, C. (2019, October 21). *Financing adaptation to climate change – an introduction (Adaptation Briefings).* Adaptation Community. https://www.adaptationcommunity.net/wp-content/uploads/2019/10/2019-10_adelphi_Adaptation-Briefings_Financing-Adaptation_an-Introduction.pdf

Cabanero, A., Nolting, L., & Praktiknjo, A. (2020). Mini-Grids for the Sustainable Electrification of Rural Areas in Sub-Saharan Africa: Assessing the Potential of KeyMaker Models. *Energies, 13*(23), 6350. doi:10.3390/en13236350

Cabré, M. M., & Araujo, J. V. (2022). *Considerations for a just and Equitable Energy Transition.* SEI (Stockholm Environment Institute). https://www.sei.org/publications/just-equitable-energy-transition/ (Accessed: 31 July 2023).

Central Intelligence Agency (CIA). (2020). Country profile: Nigeria. *CIA World Factbook.* CIA. https://www.cia.gov/library/publications/the-world-factbook/geos/sf.html

Céspedes-lorente, J. J., & Magán-díaz, A. (2016, April). (2018). Information technologies and downsizing: Examining their impact on economic performance. *Information & Management*. doi:10.1016/j.im.2018.09.012

Challinor, A., Wheeler, T., Garforth, C., Craufurd, P., & Kassam, A. (2007). Assessing the vulnerability of food crop systems in Africa to climate change. *Climatic Change*, *83*(3), 381–399. doi:10.100710584-007-9249-0

Chen, B., & Feng, Y. (1996). Some political determinants of economic growth: Theory and empirical implications. *European Journal of Political Economy*, *12*(4), 609–627. doi:10.1016/S0176-2680(96)00019-5

Cheng, C., Ren, X., Dong, K., Dong, X., & Wang, Z. (2021). How Does Technological Innovation Mitigate CO2 Emissions in OECD Countries? Heterogeneous Analysis Using Panel Quantile Regression. *Journal of Environmental Management*, *280*, 111818. doi:10.1016/j.jenvman.2020.111818 PMID:33360390

Chen, S., Saud, S., Bano, S., & Haseeb, A. (2019). The nexus between financial development, globalization, and environmental degradation: Fresh evidence from Central and Eastern European Countries. *Environmental Science and Pollution Research International*, *26*(24), 24733–24747. doi:10.100711356-019-05714-w PMID:31240660

Chin, W. (2023). *War, Technology and the State*. Bristol University Press.

CIA. (n.d.). *Sao Tome and Principe - The World Factbook*. CIA. https://www.cia.gov/the-world-factbook/countries/sao-tome-and-principe/

Coleman, S. (2005). New Mediation and Direct Representation: Reconceptualizing Representation in the Digital Age. *New Media & Society*, *7*(2), 177–198. doi:10.1177/1461444805050745

Consensys. (2018). *Welcome to the Future of Energy- Grid+ White paper*.

Coondoo, D., & Dinda, S. (2002). Causality between income and emission: A country group-specific econometric analysis. *Ecological Economics*, *40*(3), 351–367. doi:10.1016/S0921-8009(01)00280-4

Cooper, P. J. M., Stern, R. D., Noguer, M., & Gathenya, J. M. (2013). Climate change adaptation strategies in Sub-Saharan Africa: foundations for the future. In Climate change—realities, impacts over ice cap, sea level and risks. Intech Open. doi:10.5772/55133

Corfee-Morlot, J., Parks, P., Ogunleye, J., & Ayeni, F. (2022). *Achieving clean energy access in sub-Saharan Africa*. OECD. https://www.oecd.org/environment/cc/climate-futures/Achieving-clean-energy-access-Sub-Saharan-Africa.pdf

Costantini, V., & Martini, C. (2010). The causality between energy consumption and economic growth: A multi-sectoral analysis using non-stationary cointegrated panel data. *Energy Economics*, *32*(3), 591–603. doi:10.1016/j.eneco.2009.09.013

CPI. (2017). *Global Landscape of Climate Finance 2017.* CPI. https://climatepolicyinitiative.org/wpcontent/uploads/2017/10/2017-Global-Landscape-of-Climate-Finance.pdf

CPI. (2018). *Global Climate Finance: An Updated View 2018.* CPI. https://climatepolicyinitiative.org/wp-content/uploads/2018/11/Global-Climate-Finance-AnUpdated-View-2018.pdf

CPI. (2018a). *Understanding and Increasing Finance for Climate Adaptation in Developing Countries.* CPI. https://climatepolicyinitiative.org/wp-content/uploads/2018/12/Understanding-and-IncreasingFinance-for-Climate-Adaptation-in-Developing-Countries-1.pdf

Craig, O. (2018). *Concentrating Solar Power (CSP) technology adoption in South Africa.* [PhD. Thesis. Faculty of Engineering. Stellenbosch University, South Africa].

Culas, R. J. (2007). Deforestation and the environmental Kuznets curve: An institutional perspective. *Ecological Economics*, *61*(2-3), 429–437. doi:10.1016/j.ecolecon.2006.03.014

De Mesquita, B. B., Morrow, J. D., Siverson, R. M., & Smith, A. (2002). Political institutions, policy choice and the survival of leaders. *British Journal of Political Science*, *32*(4), 559–590. doi:10.1017/S0007123402000236

De, M., Mata, E., Scholten, D., & Smith, K. (2019). The multi-speed energy transition in Europe : Opportunities and challenges for EU energy security. *Energy Strategy Reviews, 26*(August). doi:10.1016/j.esr.2019.100415

DeCesaro, J. (2021). *Energy Transitions Initiative*. Office of Energy Efficiency & RE. https://www.energy.gov/eere/about-us/energy-transitions-initiative

Deing, J. (2019). *Peer-to-Peer Energy Trading Still Looks Like a Distant Prospect*. Greentech Media. https://www.greentechmedia.com/articles/read/peer-to-peer-energy-trading-still-looks-like-distant-prospect

Desai, R. M., Olofsgård, A., & Yousef, T. M. (2005). Inflation and inequality: Does political structure matter? *Economics Letters*, *87*(1), 41–46. doi:10.1016/j.econlet.2004.08.012

Diao, X. D., Zeng, S. X., Tam, C. M., & Tam, V. W. (2009). EKC analysis for studying economic growth and environmental quality: A case study in China. *Journal of Cleaner Production*, *17*(5), 541–548. doi:10.1016/j.jclepro.2008.09.007

Diestelmeier, L. (2019). Changing power: Shifting the role of electricity consumers with blockchain technology – Policy implications for EU electricity law. *Energy Policy, 128*, 189–196. doi:10.1016/j.enpol.2018.12.065

Dinda, S. (2004). Environmental Kuznets curve hypothesis: A survey. *Ecological Economics, 49*(4), 431–455. doi:10.1016/j.ecolecon.2004.02.011

Dodman, D., Ayers, J., & Huq, S. (2009). *Building Resilience in 2009 State of the World; into a Warming World.* The Worldwatch Institute.

Driessen, P. (2009). The real climate change morality crisis, Energy & Environment, 20(5), 765.

Dyson, G. (2020). The South African Power Sector: Energy Insight. *Energy and Economic Growth.* https://energyeconomicgrowth.org/sites/eeg.opml.co.uk/

Ebele, N. E., & Emodi, N. V. (2016). Climate change and its impact in Nigerian economy. *Journal of Scientific Research and Reports, 10*(6), 1–13. doi:10.9734/JSRR/2016/25162

Ecobank. (2018). *African crypto regulation*. Ecobank.

EEP Africa. (2018). *Opportunities and challenges in the mini-grid sector in Africa: Lessons learned from the EEP portfolio*. EEP Africa.

Eleanya, N. (2021). How to get rural households out of energy poverty in Nigeria: A contingent valuation. *Energy Policy.* https://www.sciencedirect.com/science/article/abs/pii/S0301421520307837

Electricity production from renewable sources, excluding hydroelectric (% of total) - Sub-Saharan Africa . (2022). World Bank. https://data.worldbank.org/indicator/EG.ELC.RNWX.ZS?locations=ZG

Electricity production from renewable, excluding hydroelectric (% of total) - Sub-Saharan Africa | Data . (2022). World Bank. https://data.worldbank.org/indicator/EG.ELC.RNWX.ZS?locations=ZG

Emodi, N. V., Emodi, C. C., Murthy, G. P., & Emodi, A. S. A. (2017). Energy policy for low carbon development in Nigeria: A LEAP model application. *Renewable & Sustainable Energy Reviews, 68*, 247–261. doi:10.1016/j.rser.2016.09.118

Erumban, A. A., & Kusum, D. (2016). Information and communication technology and economic growth in India. *Telecommunications Policy, 40*(5), 412–431. doi:10.1016/j.telpol.2015.08.006

ESMAP. (2019). *Mini Grids for Half a Billion People: Market Outlook and Handbook for Decision Makers. Executive Summary*. ESMAP.

European Bank for Reconstruction and Development. (2014, September). *Joint Report on MDB climate Finance.* EBRD. https://www.ebrd.com/downloads/news/mdb-climate-finance-2013.pdf

European Western Balkans. (2021, November 28). *Environmental protests and roadblocks across Serbia, masked men attack citizens.* European Western Balkans. https://europeanwesternbalkans.com/2021/11/28/environmental-protests-androadblocks-across-serbia-masked-men-attack-citizens/

Falobi, E. O. (2019). The Role of Renewables in Nigeria's Energy Policy Mix. *IAEE Energy Forum*, (PP. 41-46). IEEE.

Falola, T. (2018). *A History of Nigeria.* Routledge.

Farzin, Y. H., & Bond, C. A. (2006). Democracy and environmental quality. *Journal of Development Economics*, *81*(1), 213–235. doi:10.1016/j.jdeveco.2005.04.003

Fashagba, J. Y. (2009). Legislative Oversight under the Nigerian Presidential System. *Journal of Legislative Studies*, *15*(4), 439–459. doi:10.1080/13572330903302497

Fazey, I., Gamarra, J. G., Fischer, J., Reed, M. S., Stringer, L. C., & Christie, M. (2010). Adaptation strategies for reducing vulnerability to future environmental change. *Frontiers in Ecology and the Environment*, *8*(8), 414–422. doi:10.1890/080215

Federal Government of Nigeria (FGN). (2023). *Electricity Act 2023.* FGN.

Federal Ministry of Environment. (2020). *Third National Communication (TNC) of the Federal Republic of Nigeria: United Nations Framework Convention on Climate Change* (UNFCCC). FME. https://www4.unfccc.int/sites/SubmissionsStaging/NationalReports/Documents/187563_Nigeria-NC3-1-TNC%20NIGERIA%20-%2018-04-2020%20-%20FINAL.pdf

Federal Ministry of Power (FMP). (2016). *Sustainable Energy for All Action Agenda (SE4ALL-AA).* Federal Ministry of Power.

Ferguson, K. K., Soutter, L., & Neubert, M. (2019). Digital payments in Africa - how demand, technology, and regulation disrupt digital payment systems. *International Journal of Teaching and Case Studies*, *10*(4), 319. doi:10.1504/IJTCS.2019.103771

FMP. (2022). *Ministerial Performance Report: Implementation of identified priorities and deliverables of Federal Ministry of Power.* FMP.

Fonseca, C. D. (2020). *Brazilian industrialisation: Notes on the historiographical debate.* Scielo. https://www.scielo.br/scielo.php?pid

Food and Agriculture Organisation. (2018). *Analysis and Systematization on Intended Nationally Determined Contributions in Latin America and Caribbean Countries.* FAO.

Ford, J. D., Berrang-Ford, L., Biesbroek, R., Araos, M., Austin, S. E., & Lesnikowski, A. (2015). Adaptation tracking for a post-2015 climate agreement. *Nature Climate Change*, *5*(11), 967–969. doi:10.1038/nclimate2744

Frederking, L. C. (2002). Is there an endogenous relationship between culture and economic development? *Journal of Economic Behavior & Organization*, *48*(2), 105–126. doi:10.1016/S0167-2681(01)00228-1

Freedom House Organization. (2020). *Freedom House: a history*. FHO. http://www. freedomhouse. org/template. cfm

García-Quismondo, E., Santos, C., Lado, J., Palma, J., & Anderson, M. A. (2013). Optimizing the Energy Efficiency of Capacitive Deionization Reactors Working under Real-World Conditions. *Environmental Science & Technology*, *47*(20), 11866–11872. doi:10.1021/es4021603 PMID:24015835

GDP growth (annual %) - Sub-Saharan Africa. (2022). World Bank. https://data. worldbank.org/indicator/NY.GDP.MKTP.KD.ZG?locations=ZG

GEF. (2019). *GEF-7 Africa Minigrids Program*. GEF.

Geothermal Development Company. (2019). *Geothermal Development Company*. GDC. https://www.gdc.co.ke

Global Carbon Project. (2021). *Per capita CO2 emissions worldwide in 2020, by country (in metric tons)* [Graph]. In Statista. https://www.statista.com/ statistics/270508/co2-emissions-per-capita-by-country/

Global E. Y. (2019). *How blockchain can help to tackle sub-Saharan Africa's challenges*. (n.d.). EY. https://www.ey.com/en_sa/digital/tackling-sub-saharan-africa-s-challenges-with-blockchain

Global Fire Power. (2019). *Comparison Results: Military comparison results showcasing Nigeria and South Africa in side-by-side format*. Global Firepower. https://www.globalfirepower.com/countries-comparison-detail.asp?form=form& country1=nigeria&country2=south-africa&Submit=COMPARE

Glossary. (2001). *Climate Change. Education Center – Arctic Climatology and Meteorology*. NSIDC National Snow and Ice Data Center.

GOGLA. (2019). *Global Off-Grid Solar Market Report Semi-Annual Sales and Impact Data*. Gogla.

Gordon, P. (2019, July 16). Shell buys into LO3 Energy blockchain-based community platform. *Smart Energy News*. https://www.smart-energy.com/news/shell-buys-into-blockchain-based-community-energy-platform-lo3-energy/

Griebenow, C., & Ohara, A. (2019). *Report on the Brazilian Power System*. Agora Energiewende. https://www.agoraenergiewende.de/fileadmin/projekte/2019/brazil_country_profile/155_countryprof_brazil_en_web.pdf

Guillaumont, P., & Simonet, C. (2011). To What Extent Are African Countries Vulnerable to Climate Change? Lessons from a New Indicator of Physical Vulnerability to Climate Change. *Fondation pour les études et recherches sur le développement international, 5*. https://ferdi.fr/dl/df-rKATnzmJv2KH9SKi8eijFqK7/ferdi-i08-to-what-extent-are-african-countries-vulnerable-to-climate-change.pdf

Habib, S. L., Idris, N. A., Ladan, M. J., & Mohammad, A. G. (2012). Unlocking Nigeria's Solar PV and CSP Potentials for Sustainable Electricity Development. *International Journal of Scientific and Engineering Research, 3*(5), 1–8.

Hache, E. (2018). Do renewable energies improve energy security in the long run? *Inter Economics, 156*, 127–135. doi:10.1016/j.inteco.2018.01.005

Haddoum, S., Bennour, H., & Ahmed Zaïd, T. (2018). Algerian Energy Policy: Perspectives, Barriers, and Missed Opportunities. *Global Challenges (Hoboken, NJ), 2*(8), 1700134. doi:10.1002/gch2.201700134 PMID:31565341

Haider, H. (2019). Climate change in Nigeria: Impacts and responses. K4D Helpdesk Report 675. Brighton, UK: Institute of Development Studies.

Hart, D. (2020). *Deployment of Solar Photovoltaic Generation Capacity in the United States*. US Department of Energy.https://www.energy.gov/sites/prod/files/f34/Deployment%

Hartwig, J. (2010). Is health capital formation good for long-term economic growth?–Panel Granger-causality evidence for OECD countries. *Journal of Macroeconomics, 32*(1), 314–325. doi:10.1016/j.jmacro.2009.06.003

Heller, P., Harilal, K. N., & Chaudhuri, S. (2007). Building local democracy: Evaluating the impact of decentralization in Kerala, India. *World Development, 35*(4), 626–648. doi:10.1016/j.worlddev.2006.07.001

Hoffmann, R., Lee, C. G., Ramasamy, B., & Yeung, M. (2005). FDI and pollution: A granger causality test using panel data. Journal of International Development. *The Journal of the Development Studies Association, 17*(3), 311–317.

Holtz-Eakin, D., Newey, W., & Rosen, H. S. (1988). Estimating vector autoregressions with panel data. *Econometrica, 56*(6), 1371–1395. doi:10.2307/1913103

Hsiao, C. (1982). Autoregressive modeling and causal ordering of economic variables. *Journal of Economic Dynamics & Control, 4*, 243–259. doi:10.1016/0165-1889(82)90015-X

Hsiao, F. S., & Hsiao, M. C. W. (2006). FDI, exports, and GDP in East and Southeast Asia—Panel data versus time-series causality analyses. *Journal of Asian Economics, 17*(6), 1082–1106. doi:10.1016/j.asieco.2006.09.011

IEA, IRENA, UNSD, World Bank, & WHO. (2020). *Tracking SDG 7: The Energy Progress Report 2020*. WHO.

IEA. (2017). Digitalization & Energy. In Digitalization & Energy. IEA. doi:10.1787/9789264286276-en

IEA. (2018). *Market Report Series: Renewables 2018*. IEA. https://www.iea.org/reports/renewables-2018

IEA. (2019). *Africa Energy Outlook 2019 Africa Energy Outlook 2019*. IEA.

IEA. (2019). Africa Energy Outlook 2019. *World Energy Outlook Special Report*, 288.

IEA. (2021). *Renewable electricity generation increase by technology, country and region*. IEA, Paris. https://www.iea.org/data-and-statistics/charts/renewable-electricity-generation-increase-by-technology-country-and-region-2020-2021

Ihuah, P. W., & Benebo, A. M. (2014). An assessment of the causes and effects of abandonment of development projects on real property values in Nigeria. *International Journal of Research in Applied, Natural and social sciences, 2*(5), 25-36.

Imam, M. I. (2018). *Power Sector Reform and Corruption: Evidence from Sub-Saharan Africa*. EPRG. https://www.eprg.group.cam.ac.uk/wpcontent/uploads/2018/01/1801-Text.pdf

Inês, C., Guilherme, P. L., Esther, M. G., Swantje, G., Stephen, H., & Lars, H. (2020). Regulatory challenges and opportunities for collective renewable energy prosumers in the EU. *Energy Policy, 138*(April), 111212. doi:10.1016/j.enpol.2019.111212

International Hydropower Association. (2022). *Hydropower Status Report*. IHA. https://www.hydropower.org/country-profiles/nigeria

International Institute for Environment and Development (IIED). *Renewable Energy Potential in Nigeria*. IIED. https://www.iied.org/sites/default/files/pdfs/migrate/G03512.pdf

IRENA. (2012). Renewable Energy Technologies: Cost Analysis Series. Abu Dhabi: International Renewable Energy Agency (IRENA).

IRENA. (2015). [*A Roadmap for Renewable Energy Future.*]. IRENA.

IRENA. (2017). Accelerating Off-grid Renewable Energy. *Key Findings and Recommendations from IOREC*. IRENA.

IRENA. (2018). *Policies and regulations for renewable mini-grids*. International Renewable Energy Agency.

IRENA. (2019). *A new world: the geopolitics of the energy transformation*. IRENA.

IRENA. (2019). *Innovation landscape brief: Blockchain*. International Renewable Energy Agency.

IRENA. (2020). *Peer-to-peer electricity trading: Innovation Landscape Brief*. International Renewable Energy Agency. www.irena.org

IRENA. (2020). Renewable Capacity Statistics. Abu Dhabi: International Renewable Energy Agency (IRENA).

IRENA. (2020a). *Insights on Renewables: Installed Capacity Trends*. IRENA. https://www.irena.org/

IRENA. (2020b). *West Africa Clean Energy Corridor*. IRENA. https://www.irena.org/cleanenergycorridors/West-Africa-Clean-Energy-Corridor

IRENA. (2023). Renewable Energy Statistics 2023. Abu Dhabi: International Renewable Energy Agency (IRENA).

IRENA. (2023). *SIDS Lighthouses Initiative Annual Progress Report*. IRENA. https://mc-cd8320d4-36a1-40ac-83cc-3389-cdn-endpoint.azureedge.net/-/media/Files/IRENA/Agency/Publication/2023/May/IRENA_SIDS_LHI_progress_2023.pdf?rev=6aac8f77eede4b768a078cc4a971c543

ITU. (2015). Focus Group on Smart Sustainable Cities. *Information and communication technologies for climate change adaptation in cities*. UNFCCC. https://www4.unfccc.int/sites/NAPC/Documents/Supplements/ICTs-for-climate-change-adaptation.pdf

Jardini, J. A., Ramos, D., Martini, J., Reis, L., & Tahan, C. (2012). Brazilian Energy Crisis. *Power Engineering Review*, *22*(2), 1–24.

Jimoh, M. A., & Raji, B. S. (2023). Electric Grid Reliability: An Assessment of the Nigerian Power System Failures, Causes and Mitigations. *Covenant Journal of Engineering Technology*, *7*(1), 17–20.

Jolliffe, I. (2005). *Principal component analysis*. Encyclopedia of statistics in behavioral science.

Jordan, E. A., Kusakana, K., & Bokopane, L. (2018). Prospective Architecture for Local Energy Generation and Distribution with Peer-to-Peer Electricity Sharing in a South African Context. *2018 Open Innovations Conference, OI2018*, (pp. 161–164). IEEE. 10.1109/OI.2018.8535971

Jorgenson, D. W., Ho, M. S., & Samuels, J. D. (2016). The impact of information technology on postwar US economic growth. *Telecommunications Policy*, *40*(5), 398–411. doi:10.1016/j.telpol.2015.03.001

Joseph, J. (2001). Sustainable development and democracy in the megacities. *Development in Practice*, *11*(2-3), 218–231. doi:10.1080/09614520120056360

Justesen, M. K. (2008). The effect of economic freedom on growth revisited: New evidence on causality from a panel of countries 1970–1999. *European Journal of Political Economy*, *24*(3), 642–660. doi:10.1016/j.ejpoleco.2008.06.003

Kaufmann, D., Kraay, A., & Mastruzzi, M. (2009). Governance matters VIII: Aggregate and individual governance indicators, 1996-2008. *World Bank Policy Research Working Paper*, 4978. doi:10.1596/1813-9450-4978

Khan, S. U. D. (2020). Environmental sustainability: a clean energy aspect versus poverty. *Environmental Science and Pollution Research, 28*(11), 13097–13104. doi:10.1007/s11356-020-11520-6

Khatoon, A., Verma, P., Southernwood, J., Massey, B., & Corcoran, P. (2019). Blockchain in energy efficiency: Potential applications and benefits. *Energies*, *12*(17), 1–14. doi:10.3390/en12173317

Kiyasseh, L. (2022). *Strong momentum in Saudi Arabia's drive toward renewables and infrastructure.* Middle East Institute. https://www.mei.edu/publications/strong-momentum-saudi-arabias-drive-toward-renewables-and-infrastructure

Klein, L. P., Krivoglazova, A., Matos, L., Landeck, J., & de Azevedo, M. (2019). A novel peer-to-peer energy sharing business model for the Portuguese energy market. *Energies*, *13*(1), 125. doi:10.3390/en13010125

Koffman, T. (2019). *Blockchain - Africa Rising*. Forbes. https://www.forbes.com/sites/tatianakoffman/2019/04/04/blockchain-africa-rising/#6e22abc7711f

Kónya, L. (2006). Exports and growth: Granger causality analysis on OECD countries with a panel data approach. *Economic Modelling*, *23*(6), 978–992. doi:10.1016/j.econmod.2006.04.008

KPMG. (2019). *Nigeria ranks 131 in World Bank's 2020 Doing Business Report.* KPMG. https://kpmg.com/ng/en/home/insights/2019/10/nigeria-ranks-131-in-world-bank-s-2020-doing-business-report.html

Le, H., & Baboyan, K. (2015). *Climate Budget Tagging County – Driven Initiative in Tracking Climate Change Expenditure: The Case of Bangladesh, Indonesia, Nepal and Philippines.* (UNDP Working Paper). UNDP.

Leandro, F. J. B. S., Martínez-Galán, E., & Gonçalves, P. (2023). *Portuguese-speaking Small Island Developing States*. Springer Nature Singapore. doi:10.1007/978-981-99-3382-2

Lee, C. C., Chang, C. P., & Chen, P. F. (2008). Energy-income causality in OECD countries revisited: The key role of capital stock. *Energy Economics*, *30*(5), 2359–2373. doi:10.1016/j.eneco.2008.01.005

Lightency. (2020). *Lightency-The Bright Side of Energy*. Lightency. http://lightency.io/#/

Lim, B., Spanger-Siegfried, E., Burton, I., Malone, E., & Huq, S. (2005). *Adaptation policy frameworks for climate change: developing strategies, policies and measures*. United Nations Development Programme.

Lin, B., & Chen, Y. (2019). Does electricity price matter for innovation in renewable energy technologies in China? *Energy Economics*, *78*, 259–266. doi:10.1016/j.eneco.2018.11.014

Lobell, D. B. (2014). Climate change adaptation in crop production: Beware of illusions. *Global Food Security*, *3*(2), 72–76. doi:10.1016/j.gfs.2014.05.002

Lobell, D. B., Burke, M. B., Tebaldi, C., Mastrandrea, M. D., Falcon, W. P., & Naylor, R. L. (2008). Prioritizing climate change adaptation needs for food security in 2030. *Science*, *319*(5863), 607–610. doi:10.1126cience.1152339 PMID:18239122

Ma, Z., Bloch-Hansen, K., Buck, J. W., Hansen, A. K., Henriksen, L. J., Thielsen, C. F., Santos, A. Q., & Jorgensen, B. N. (2018). Peer-to-Peer Trading Solution for Microgrids in Kenya. *IEEE PES/IAS PowerAfrica*, (pp. 420–425). IEEE. doi:10.1109/PowerAfrica.2018.8520980

Macrotrends (2020). Nigeria Literacy Rate 1991-2020. *MacroTrends*. https://www.macrotrends.net/countries/NGA/nigeria/literacy-rate

Magnani, E. (2001). The Environmental Kuznets Curve: Development path or policy result? *Environmental Modelling & Software*, *16*(2), 157–165. doi:10.1016/S1364-8152(00)00079-7

Mahlalela, P. T., Blamey, R. C., Hart, N. C. G., & Reason, C. J. C. (2020). Drought in the Eastern Cape region of South Africa and trends in rainfall characteristics. *Climate Dynamics*, *55*(9–10), 2743–2759. doi:10.100700382-020-05413-0 PMID:32836893

Maizland, L. (2021). *Global Climate Agreements: Successes and Failures.* Council on Foreign Relations. www.cfr.org. https://www.cfr.org/backgrounder/paris-global-climate-change-agreements

Majeed Butt, O., Zulqarnain, M., & Majeed Butt, T. (2021). Recent advancement in smart grid technology: Future prospects in the electrical power network. *Ain Shams Engineering Journal*, *12*(1), 687–695. doi:10.1016/j.asej.2020.05.004

Malik, S. (2019). *Africa's Wind Project Pipeline Grows to 18GW.* Greentech Media. https://www.greentechmedia.com/articles/read/africa-18-gw-wind-project-pipeline

Mallowah, S., & Oyier, C. (2022). The Environment and Climate Change Law Review. *The Law Reviews*. thelawreviews.co.uk. https://thelawreviews.co.uk/title/the-environment-and-climate-change-law-review/kenya#:~:text=i%20The%20Constitution%20of%20Kenya%202010&text=Article%2070%20reinforces%20the%20right,Land%20Court%20under%20Article%20162

Marigi, S. N. (2017). Climate Change Vulnerability and Impacts Analysis in Kenya. *American Journal of Climate Change*, *6*(1), 54–72. doi:10.4236/ajcc.2017.61004

McKenzie, B. (2018). *Blockchain and Cryptocurrency in Africa.*

Menegat, R. (2002). Participatory democracy and sustainable development: Integrated urban environmental management in Porto Alegre, Brazil. *Environment and Urbanization*, *14*(2), 181–206. doi:10.1177/095624780201400215

Mengelkamp, E., Gärttner, J., Rock, K., Kessler, S., Orsini, L., & Weinhardt, C. (2017). Designing microgrid energy markets: A case study: The Brooklyn Microgrid. *Applied Energy*, *210*, 870–880. doi:10.1016/j.apenergy.2017.06.054

Micale, V., Tonkonogy, B., & Mazza, F. (2018). *Understanding and Increasing Finance for Climate Adaptation in Developing Countries*. Climate Policy Initiatve.

Michaelowa, A., Hoch, S., Honegger, M., & Friedmann, V. (2016). *TRANSITIONING FROM INDCs TO NDCs IN AFRICA*. African Development Bank. Retrieved from African Development Bank website. https://www.afdb.org/fileadmin/uploads/afdb/Documents/Publications/AfDB-CIF-Transitioning_fromINDCs_to_NDC-report-November2016.pdf

Migliaccio, G. (2016). ICT for Disability Management in the Net Economy. *International Journal of Globalisation and Small Business*, *8*(1), 51. doi:10.1504/IJGSB.2016.076452

Miljkovic, D., & Rimal, A. (2008). The impact of socio-economic factors on political instability: A cross-country analysis. *Journal of Socio-Economics*, *37*(6), 2454–2463. doi:10.1016/j.socec.2008.04.007

Ministério das Infraestruturas e Recursos Naturais. (2021). *Plano de Ação Nacional no Sector das Energias Renováveis*. Ministério das Infraestruturas e Recursos Naturais.

Miyakoshi, T., & Tsukuda, Y. (2004). The causes of the long stagnation in Japan. *Applied Financial Economics*, *14*(2), 113–120. doi:10.1080/0960310042000176380

Moradiya, M. (2019). *The Challenges Renewable Energy Sources Face*. AZoCleantech. https://www.azocleantech.com/article.aspx?ArticleID=836

Morstyn, T., Farrell, N., Darby, S. J., & McCulloch, M. D. (2018). Using peer-to-peer energy-trading platforms to incentivize prosumers to form federated power plants. In Nature Energy (Vol. 3, pp. 94–101). Nature Publishing Group. doi:10.103841560-017-0075-y

Moser, S. C., & Ekstrom, J. A. (2010). A framework to diagnose barriers to climate change adaptation. *Proceedings of the National Academy of Sciences of the United States of America*, *107*(51), 22026–22031. doi:10.1073/pnas.1007887107 PMID:21135232

Mrabure, K. O. (2022). Nuclear Power as a Source of Energy in Nigeria and Sustainability of the Environment, NIALS. *Journal of Environmental Law*, *5*, 235–275.

Mukheibir, P., Kuruppu, N., Gero, A., & Herriman, J. (2013). Overcoming cross-scale challenges to climate change adaptation for local government: A focus on Australia. *Climatic Change*, *121*(2), 271–283. doi:10.100710584-013-0880-7

Munsing, E., Mather, J., & Moura, S. (2017). Blockchains for decentralized optimization of energy resources in microgrid networks. *IEEE Conference on Control Technology and Applications (CCTA), 2017-Janua*, (pp. 2164–2171). IEEE. 10.1109/CCTA.2017.8062773

Mylrea, M., Nikhil, S., & Gourisetti, G. (2017). *Blockchain for Smart Grid Resilience: Exchanging Distributed Energy at Speed.* Scale and Security.

Narayan, P. K., Nielsen, I., & Smyth, R. (2008). Panel data, cointegration, causality and Wagner's law: Empirical evidence from Chinese provinces. *China Economic Review*, *19*(2), 297–307. doi:10.1016/j.chieco.2006.11.004

Ndung'u, N., & Signé, L. (2020). The Fourth Industrial Revolution and digitization will transform Africa into a global powerhouse. In Capturing the Fourth Industrial Revolution: A regional and national agenda (pp. 61–73). Foresight Africa 2020 report.

Net migration - Sub-Saharan Africa . (2022). World Bank. https://data.worldbank.org/indicator/SM.POP.NETM?locations=ZG

New, M., Liverman, D., Schroder, H., & Anderson, K. (2011). Four degrees and beyond: the potential for a global temperature increase of four degrees and its implications: Introduction. *Philosophical Transactions of the Royal Society A: Mathematical, Physical and Engineering Sciences, 369*(1934), 6-19.

Newman, B. A., & Thomson, R. J. (1989). Economic growth and social development: A longitudinal analysis of causal priority. *World Development*, *17*(4), 461–471. doi:10.1016/0305-750X(89)90255-6

Nick, M. (2020). *Energy and GDP in Nigeria.* Stanford. http://large.stanford.edu/courses/2016/ph240/nick2/

Nieman, A. (2015). *A FEW SOUTH AFRICAN CENTS ' WORTH ON BITCOIN. 18*(5).

NIMET (2008). *Nigeria Climate Review Bulletin 2007.* Nigerian Meteorological Agency.

Niyibizi, A., & Komakech, A. (2013). Climate Change Mitigation in Developing Countries Using ICT as an Enabling Tool. In W. Leal Filho, F. Mannke, R. Mohee, V. Schulte, & D. Surroop (Eds.), *Climate-Smart Technologies. Climate Change Management*. Springer. doi:10.1007/978-3-642-37753-2_2

Nkiriki, J., & Ustun, T. S. (2017). Mini-grid policy directions for decentralized smart energy models in Sub-Saharan Africa. *2017 IEEE PES Innovative Smart Grid Technologies Conference Europe, ISGT-Europe 2017 - Proceedings, 2018-Janua*, (pp. 1–6). 10.1109/ISGTEurope.2017.8260217

Noor, S., Yang, W., Guo, M., van Dam, K. H., & Wang, X. (2018). Energy Demand Side Management within micro-grid networks enhanced by blockchain. *Applied Energy, 228*, 1385–1398. doi:10.1016/j.apenergy.2018.07.012

Norouzi, N., & Dehghani, M. A. (2020). A Backward Scenario Planning Overview of the Greenhouse Gas Emission in Iran by the End of the Sixth Progress Plan. *Current Environmental Management (Formerly: Current Environmental Engineering), 7*(1), 13-35.

Norouzi, N. (2019). An overview on Water, Energy & Environment by 2030. *International Journal of Management Perspective, 8*(2), 11–19.

Norouzi, N. (2020). Climate change impacts on the water flow to the reservoir of the Dez Dam basin. *Water Cycle, 1*, 113–120. doi:10.1016/j.watcyc.2020.08.001

Norouzi, N., & Fani, M. (2020). Black gold falls, black plague arise-An Opec crude oil price forecast using a gray prediction model. *Upstream Oil and Gas Technology, 5*, 100015. doi:10.1016/j.upstre.2020.100015

Norouzi, N., & Fani, M. (2021a). The seventh line: A scenario planning strategic framework for Iranian 7th energy progress plan by 2020-2025. *Journal of Energy Management and Technology, 5*(3), 43–53.

Norouzi, N., & Fani, M. (2021b). The prioritization and feasibility study over renewable technologies using fuzzy logic: A case study for Takestan plains. *Journal of Energy Management and Technology, 5*(2), 12–22.

Norouzi, N., Fani, M., & Ziarani, Z. K. (2020). The fall of oil Age: A scenario planning approach over the last peak oil of human history by 2040. *Journal of Petroleum Science Engineering, 188*, 106827. doi:10.1016/j.petrol.2019.106827

Norouzi, N., & Kalantari, G. (2020). The sun food-water-energy nexus governance model a case study for Iran. *Water-Energy Nexus*, *3*, 72–80. doi:10.1016/j.wen.2020.05.005

Nosiri, U. D., & Ohazurike, E. U. (2016). Border Security and National Security in Nigeria. *Southeast Journal of Political Science*, *2*(2), 214–226.

NTEP. (2022). *Nigeria's pathway to achieve carbon neutrality by 2060*. Nigeria Energy Transition Plan.

Nwankwo, O. C., & Njogo, B. O. (2013). The Effect of Electricity Supply on Industrial Production Within the Nigerian Economy (1970 – 2010). *Journal of Energy Technologies and Policy*, *3*(4), 34–42.

Nyoka, N. (2022). *Inga 3 hydroelectric scheme is a looming disaster*. New Frame. https://www.newframe.com/inga-3-hydroelectric-scheme-is-a-looming-disaster/

Obani, P. C., & Gupta, J. (2016). The impact of economic recession on climate change: Eight trends. *Climate and Development*, *8*(3), 211–223. doi:10.1080/17565529.2015.1034226

Obiora, C. A., Chiamogu, A. P., & Chiamogu, U. P. (2019). *Power Devolution and Electricity Transmission in Nigeria: A Study in Resources Mobilization for Economic Development*. Advances in Social Sciences Research Journal.

ODI. (2017). *Climate Finance Thematic Briefing: Adaptation Finance*. ODI. https://www.odi.org/sites/odi.org.uk/files/resource-documents/12073.pdf

OECD. (2008). *Handbook on constructing composite indicators: methodology and user guide*. OECD.

OECD. (2019). The Policy Environment for Blockchain Innovation and Adoption: 2019 OECD Global Blockchain Policy Forum Summary Report. *OECD Blockchain Policy Series*. OECD. www.oecd.org/finance/2019-OECD-Global-Blockchain-Policy-Forum-Summary-Report.pdf

OECD. DAC. (2011). Handbook on the OECD-DAC Climate markers. *Paris: Organisation for Economic Co-operation and Development's Development Assistance Committee (DAC).* OECD. https://www. oecd. org/dac/stats/48785310. pdf

Off Grid Energy Independence. (2019). *ENGIE acquires Mobisol, becomes market leader in off-grid solar Africa*. Off Grid Energy Independence. https://www.offgridenergyindependence.com/articles/18086/engie-acquires-mobisol-becomes-market-leader-in-off-grid-solar-africa

Ogunbiyi, D. (2020, July). *Nigeria is using the pandemic to build a better energy future.* World Economic Forum. https://www.weforum.org/agenda/2020/09/nigeria-using-pandemic-build-sustainable-energy-future/

Ogundipe, A. A., & Akinyemi, O. (2013). Electricity Consumption and Economic Development in Nigeria. *Journal of Business Management and Applied Economics*, *2*(4), 1–13.

Ogunleye, O. A. (2021). *Techno-Economic Analysis of Concentrated Solar Power Projects in Northern Nigeria.* University of Ibadan Postgraduate College. https://pgsds.ictp.it/xmlui/handle/123456789/1258

Ogunlowo, O. O., Bristow, A. L., & Sohail, M. (2015). Developing compressed natural gas as an automotive fuel in Nigeria: Lessons from international markets. *Energy Policy*, *76*, 7–17. doi:10.1016/j.enpol.2014.10.025

Ogunmodimu, O. (2012). *Potential Contribution of Solar Thermal Power to Electricity Supply in Northern Nigeria* [Thesis, University of Cape Town].

Ogunmodimu, O., & Okoroigwe, E. C. (2019). Solar thermal electricity in Nigeria: Prospects and challenges. Energy Policy. *Renewable Energy*, *128*, 440–448.

Ohajianya, A., Abumere, O., Owate, I., & Osarolube, E. (2014). Erratic Power Supply in Nigeria: Causes and Solutions. *International Journal of Engineering Science Invention*, *3*, 51–55.

Ohunakin, O. S., Matthew, O. J., Adaramola, S., Atiba, O. E., Adelekan, D. S., Aluko, O. O., Henry, E. U., & Ezekiel, V. U. (2023). Techno-economic assessment of offshore wind energy potential at selected sites in the Gulf of Guinea. *Energy Conversion and Management*, *288*, 117–110. doi:10.1016/j.enconman.2023.117110

Oke, Y. (2021). Nigerian electricity law and practice, (2nd edition). Lagos, Nigeria, Princeton and Associates Publishing Co. Ltd.

Oke, Y. (2021). Nigerian energy resources law and practice, (2nd edition). Lagos, Nigeria, Princeton and Associates Publishing Co. Ltd.

Oliveira, T. A., Oliver, M., & Ramalhinho, H. (2020). Challenges for Connecting Citizens and Smart Cities: ICT, E-Governance and Blockchain. *Sustainability (Basel)*, *12*(7), 2926. doi:10.3390u12072926

Olorunmola, A. (2016). *Cost of politics in Nigeria*. Westminster Foundation for Democracy.

Olson, M. (1993). Dictatorship, democracy, and development. *The American Political Science Review*, *87*(3), 567–576. doi:10.2307/2938736

Omenda, P., & Teklemariam, M. (2010). *Overview Of Geothermal Resource Utilization In The East African Rift System*. RISE. https://rise.esmap.org/data/files/library/kenya/Renewable Energy/RE 9.2 Omenda Teklemariam 2010 Overview of Geothermal Resource Utilization in the East African Rift System.pdf

Omotola, J. S. (2010). Elections and democratic transition in Nigeria under the Fourth Republic. *African Affairs*, *109*(437), 535–553. doi:10.1093/afraf/adq040

Onakoya, A. (2013). Energy Consumption and Nigerian Economic Growth: An Empirical Analysis. *European Scientific Journal*, *9*(4), 25–33.

OneWattSolar. (2020). *Blockchain Powered Digital Electricity - OneWattSolar*. OneWattSolar. https://onewattsolar.com/

Onifade, O. W. F. (2020). Artificial Intelligence and National Development. University of Ibadan.

Oprea, S. V., & Bâra, A. (2021). Devising a trading mechanism with a joint price adjustment for local electricity markets using blockchain. Insights for policy makers. *Energy Policy*, *152*(February), 112237. doi:10.1016/j.enpol.2021.112237

Organization for Economic Cooperation and Development. (2011, September). *Handbook on the OECD - DAC Climate Markers.* OECD. https://www.oecd.org/dac/stats/48785310.pdf

Oseni, P. (2011). An analysis of the power sector performance in Nigeria. *Renewable & Sustainable Energy Reviews*, *15*(9), 4765–4774. doi:10.1016/j.rser.2011.07.075

Oureilidis, K., Malamaki, K. N., Gallos, K., Tsitsimelis, A., Dikaiakos, C., Gkavanoudis, S., Cvetkovic, M., Mauricio, J. M., Ortega, J. M. M., Ramos, J. L. M., Papaioannou, G., & Demoulias, C. (2020). Ancillary services market design in distribution networks: Review and identification of barriers. *Energies*, *13*(4), 917. doi:10.3390/en13040917

Oyedele, T. (2021). *Economic and fiscal implications of Nigeria's rebased GDP*. PWC. https://www.pwc.com/ng/en/publications/gross-domestic-product-does-size-really-matter.html

Oyedepo, S. O. (2012). On Energy for Sustainable Development in Nigeria. *Renewable & Sustainable Energy Reviews*, *16*(9), 2583–2598. doi:10.1016/j.rser.2012.02.010

Oyewo, A., Aghahosseini, A., Bogdanov, D., & Breyer, C. (2018). Pathways to a fully sustainable electricity supply for Nigeria in the mid-term future. *Energy Conversion and Management*, *178*(1), 44–64. doi:10.1016/j.enconman.2018.10.036

Palvia, P., Baqir, N., & Nemati, H. (2015). ICT Policies in Developing Countries: AN Evaluation with the Extended Design-Actuality Gaps Framework. *The Electronic Journal on Information Systems in Developing Countries*, *71*(1), 1–34. doi:10.1002/j.1681-4835.2015.tb00510.x

Paribas, B. N. P. (2019). *Blockchain is more than just numbers for these small farmers.* CIB. https://cib.bnpparibas.com/sustain/blockchain-is-more-than-just-numbers-for-these-small-farmers_a-3-3149.html

Pereira, M. (2018). Energy poverty and climate change elements to debate. In: Debra J. Davidsons; Matthias Gross. (Org.). Handbook on Energy and Society. Oxford University Press.

Pfeifer, A., Dobravec, V., Pavlinek, L., & Kraja, G. (2018). Integration of renewable energy and demand response technologies in interconnected energy systems. *Energy*, *161*, 447–455. doi:10.1016/j.energy.2018.07.134

Piñeiro Chousa, J., Tamazian, A., & Vadlamannati, K. C. (2017). Does higher economic and financial development lead to environmental degradation: Evidence from BRIC countries. *Energy Policy*, *37*(1), 2009.

PLAAS. (2018). *Supply Chain Research Architecture*. PLAAS.

Poverty headcount ratio at $1.90 daily (2011) PPP (% of the population) - Sub-Saharan Africa. (2022). World Bank. https://data.worldbank.org/indicator/SI.POV.DDAY?locations=ZG

Pradhan, R. P., Arvin, M. B., & Norman, N. R. (2015). The dynamics of information and communications technologies infrastructure, economic growth, and financial development: Evidence from Asian countries. *Technology in Society*, *42*, 135–149. doi:10.1016/j.techsoc.2015.04.002

Pradhan, R. P., Mallik, G., & Bagchi, T. P. (2018). Information communication technology (ICT) infrastructure and economic growth: A causality evinced by cross-country panel data. *IIMB Management Review*, *30*(1), 91–103. doi:10.1016/j.iimb.2018.01.001

PricewaterhouseCoopers. (2016). Blockchain – an opportunity for energy producers and consumers? *Pwc.Com*, 1–45.

Rahman, M., & Mezbah-ul-Islam, M. (2012). *Issues and Challenges for Sustainable Digital Preservation Practices in Bangladesh*. Research Gate. https://www.researchgate.net/publication/261178681_Issues_and_Challenges_for_Sustainable_Digital_Preservation_Practices_in_Bangladesh

Rao, P. (2018). Africa could be the next frontier for cryptocurrency. *Africa Renewal*, *32*(1), 27–27. doi:10.18356/f6b3e553-en

Raouf, M. (2008). *Climate Change Threats, Opportunities, and the GCC Countries*. Middle East Institute. https://www.mei.edu/publications/climate-change-threats-opportunities-and-gcc-countries

REA. (n.d.). *Solar Power Naija - Enabling 5 Million new Connections*. REA. https://rea.gov.ng/solar-power-naija/#:~:text=The%205Million%20Solar%20Power%20Naija,II

Renewable energy – powering a safer future. (2022). United Nations. https://www.un.org/en/climatechange/raising-ambition/renewable-energy

Renewable energy capacity worldwide by country. (2020). Statista. https://www.statista.com/statistics/267233/renewable-energy-capacity-worldwide-by-country/

Renewable energy consumption (% of total final energy consumption) - Sub-Saharan Africa. (2022). World Bank. https://data.worldbank.org/indicator/EG.FEC.RNEW.ZS?locations=ZG

Renewable energy consumption (% of total final energy consumption) - Sub-Saharan Africa.. (2022). World Bank. https://data.worldbank.org/indicator/EG.FEC.RNEW.ZS?locations=ZG

Reporters without Borders. (2020). *Press Freedom Index*. RSF. www.rsf.org

República Democrática de São Tomé e Príncipe. (2019). *Plano Nacional de Desenvolvimento Sustentável de São Tomé e Príncipe*. Republica Democratica. https://financas.gov.st/phocadownload/Planeamento/publicacao/Plano%20Nacional%20de%20Desenvolvimento%20Sustentavel%20-%20STP%20-%202020-2024.pdf

Resch, E., Allan, S., Álvarez, L. G., & Bisht, H. (2017). *Mainstreaming, accessing and institutionalising finance for climate change adaptation. ACT Learning Paper*. OPM.

Ribeiro, F., Ferreira, P., Araújo, M., & Braga, A. C. (2018). Modeling perception and attitudes towards renewable energy technologies. *Renewable Energy*, *122*, 688–697. doi:10.1016/j.renene.2018.01.104

Riti, J. S., Song, D., Shu, Y., Kamah, M., & Atabani, A. A. (2018). Does Renewable Energy Ensure Environmental Quality in Favour of Economic Growth? Empirical Evidence from China's Renewable Development. *Quality & Quantity*, *52*(5), 2007–2030. doi:10.100711135-017-0577-5

Rogger, D. (2014). The causes and consequences of political interference in bureaucratic decision making: Evidence from Nigeria. *Job Market Paper*, *12*(1), 1–22.

Rot, A., Sobińska, M., Hernes, M., and Franczyk, B. (2020). Digital Transformation of Public Administration through Blockchain Technology. *Towards Industry 4.0—Current Challenges in Information Systems*. Springer. doi:10.1007/978-3-030-40417-8_7

Rubio-Varas, M., & Muñoz-Delgado, B. (2019a). Long-term diversification paths and energy transitions in Europe. *Ecological Economics*, *163*(June), 158–168. doi:10.1016/j.ecolecon.2019.04.025

Rubio-Varas, M., & Muñoz-Delgado, B. (2019b). The Energy Mix Concentration Index (EMCI): Methodological considerations for implementation. *MethodsX*, *6*(May), 1228–1237. doi:10.1016/j.mex.2019.05.023 PMID:31193910

Rumble, O. (2018). Climate change legislative development on the African continent. *4th Scientific Conference of the Association of Environmental Law Lecturers from African Universities in cooperation with the Climate Policy and Energy Security Programme for Sub-Saharan Africa of the Konrad- Adenauer-Stiftung and UN Environment.* UN Environment.

Saberi, S., Kouhizadeh, M., Sarkis, J., & Shen, L. (2019). Blockchain Technology and its Relationships to Sustainable Supply Chain Management. *International Journal of Production Research*, *57*(7), 2117–2135. doi:10.1080/00207543.2018.1533261

Sachgau, O. (2022). *How to fight climate change now.* Siemens-Energy. https://www.siemens-energy.com/global/en/news/magazine/2022/how-to-fight-climate-change-now.html

Sachikonye, L. M. (2002). Democracy, Sustainable Development and Poverty: Are they Compatible? *Occasional Paper*, (2), 4. Development Policy Management Forum (DPMF).

Samuel, O., Almogren, A., Javaid, A., Zuair, M., Ullah, I., & Javaid, N. (2020). Leveraging blockchain technology for secure energy trading and least-cost evaluation of decentralized contributions to electrification in sub-Saharan Africa. *Entropy (Basel, Switzerland)*, *22*(2), 226. doi:10.3390/e22020226 PMID:33286000

Sarangi, G. K. (2018). *Green energy finance in India: Challenges and solutions.* Asian Development Bank. https://www.adb.org/publications/green-energy-finance-india-challenges-and-solutions

Schröder, A., Kunz, F., Meiss, J., Mendelevitch, R., & von Hirschhausen, C. (2013). Current and prospective costs of electricity generation until 2050, DIW Data Documentation, No. 68. Deutsches Institut für Wirtschaftsforschung, DIW.

Schuftan, C. (2005). Dignity Counts: A guide to using budget analysis to advance human rights. *Social Change*, *35*(1), 143–146. doi:10.1177/004908570503500113

Segreto, M., Principe, L., Desormeaux, A., Torre, M., Tomassetti, L., Tratzi, P., Paolini, V., & Petracchini, F. (2020). Trends in social acceptance of renewable energy across Europe: A literature review. *International Journal of Environmental Research and Public Health*, *17*(24), 9161. doi:10.3390/ijerph17249161 PMID:33302464

Shahbaz, M., Haouas, I., Sohag, K., & Ozturk, I. (2020). The financial development-environmental degradation nexus in the United Arab Emirates: The importance of growth, globalization and structural breaks. *Environmental Science and Pollution Research International*, *27*(10), 1–15. doi:10.100711356-019-07085-8 PMID:31950417

Shahbaz, M., Nasir, M. A., & Roubaud, D. (2018). Environmental degradation in France: The effects of FDI, financial development, and energy innovations. *Energy Economics*, *74*, 843–857. doi:10.1016/j.eneco.2018.07.020

Simões, J., Leandro, F. J., de Sousa, E. C., & Oberoi, R. (Eds.). (2023). *Changing the Paradigm of Energy Geopolitics*. Peter Lang Verlag., doi:10.3726/b18776

Smit, B., & Skinner, M. W. (2002). Adaptation options in agriculture to climate change: A typology. *Mitigation and Adaptation Strategies for Global Change*, *7*(1), 85–114. doi:10.1023/A:1015862228270

Söderbaum, P. (2007). Issues of paradigm, ideology and democracy in sustainability assessment. *Ecological Economics*, *60*(3), 613–626. doi:10.1016/j.ecolecon.2006.01.006

Solshare. (2020). *ME Solshare*. https://me-solshare.com/

Son, Y.-B., Im, J.-H., Kwon, H.-Y., Jeon, S.-Y., & Lee, M.-K. (2020). Privacy-Preserving Peer-to-Peer Energy Trading in Blockchain-Enabled Smart Grids Using Functional Encryption. *Energies*, *13*(6), 1321. doi:10.3390/en13061321

Stafford, M. S., Horrocks, L., Harvey, A., & Hamilton, C. (2011). Rethinking adaptation for a 4° C world. *Philosophical transactions. Series A, Mathematical, physical, and engineering sciences, 369*(1934), 196-216.

Stallo, C., De Sanctis, M., Ruggieri, M., Bisio, I., & Marchese, M. (2010) ICT Applications in Green and Renewable Energy Sector. *Workshops on Enabling Technologies: Infrastructure for Collaborative Enterprises.*

Statista. (2020). *Nigeria: Distribution of gross domestic product (GDP) across economic sectors from 2009 to 2019*. Statista. https://www.statista.com/statistics/382311/nigeria-gdp-distribution-across-economic-sectors/

Stern, D. I. (2004). The rise and fall of the environmental Kuznets curve. *World Development*, *32*(8), 1419–1439. doi:10.1016/j.worlddev.2004.03.004

Stern, D. I., Common, M. S., & Barbier, E. B. (1996). Economic growth and environmental degradation: The environmental Kuznets curve and sustainable development. *World Development*, *24*(7), 1151–1160. doi:10.1016/0305-750X(96)00032-0

Stigka, E. K., Paravantis, J. A., & Mihalakakou, G. K. (2014). Social acceptance of renewable energy sources: A review of contingent valuation applications. *Renewable & Sustainable Energy Reviews*, *32*, 100–106. doi:10.1016/j.rser.2013.12.026

Stirling, A. (2008). Diversity and sustainable energy transitions: Multicriteria diversity analysis of electricity portfolios. In M. Bazilian & F. Roques (Eds.), *Analytical Methods for Energy Diversity and Security* (pp. 1–29). Elsevier Ltd., doi:10.1016/B978-0-08-056887-4.00001-9

Stocker, A., Alshawish, A., Bor, M., Vidler, J., Gouglidis, A., Scott, A., Marnerides, A., De Meer, H., & Hutchison, D. (2022). An ICT architecture for enabling ancillary services in Distributed Renewable Energy Sources based on the SGAM framework. *Energy Informatics*, *5*(1), 5. doi:10.118642162-022-00189-5

The United Nations Framework Convention on Climate Change. (1994). *About.* UNFCCC. https://unfccc.int/process-and-meetings/the-convention/what-is-the-united-nations-framework-convention-on-climate-change

The World Bank. (2020). *GDP growth (annual %) - South Africa.* World Bank. https://data.worldbank.org/indicator/NY.GDP.MKTP.KD.ZG?locations=ZA

The World Bank. (2023). *Databank – Population estimates and projections.* The World Bank. https://databank.worldbank.org/source/population-estimates-and-projections

The World Bank. (2023). *GDP (Current US$) – Nigeria.* The World Bank. https://data.worldbank.org/indicator/NY.GDP.MKTP.CD?locations=NG

The World Bank. (2023). *Population total – Nigeria.* The World Bank. https://data.worldbank.org/indicator/SP.POP.TOTL?end=2022&locations=NG&start=2022&view=bar

Thiede, B., & Strube, J. (2020). Climate variability and child nutrition: Findings from sub-Saharan Africa. *Global Environmental Change*, *65*, 102192. doi:10.1016/j.gloenvcha.2020.102192 PMID:34789965

Thompson, M. (2018). Social capital, innovation, and economic growth. *Journal of Behavioral and Experimental Economics*, *73*, 46–52. doi:10.1016/j.socec.2018.01.005

Thorarinsdottir, T. L., & de Bruin, K. (2016), Challenges of climate change adaptation, *Eos, 97,* https://doi.org/ doi:10.1029/2016EO062121

Toader, E., Firtescu, B., Roman, A., & Anton, S. (2018). Impact of Information and Communication Technology Infrastructure on Economic Growth: An Empirical Assessment for the EU Countries. *Sustainability (Basel)*, *10*(10), 3750. doi:10.3390u10103750

Total greenhouse gas emissions (kt of CO2 equivalent) - Sub-Saharan Africa. (2022). World Bank. https://data.worldbank.org/indicator/EN.ATM.GHGT.KT.CE?locations=ZG

Trading Economics. (2020). *Nigeria - Economically Active Population in Agriculture*. Trading Economies. https://tradingeconomics.com/nigeria/economically-active-population-in-agriculture-number-wb-data.html

Trading Economics. (2020). *Nigeria GDP1960-2019 Data*. Trading Economics. https://tradingeconomics.com/nigeria/gdp

Trading Economics. (2020). *Nigeria Population*. Trading Economics. https://tradingeconomics.com/nigeria/population

Trading Economics. (2021). *Nigeria Corruption Index1996-2020 Data*. Trading Economics. https://tradingeconomics.com/nigeria/corruption-index

Transmission Company of Nigeria. (2021). *TCN Successfully Transmits Enhanced All-Time Peak*. TCN. https://tcn.org.ng/blog_post_sidebar104.php

Tushar, W., Saha, T. K., Yuen, C., Smith, D., & Poor, H. V. (2020). Peer-to-Peer Trading in Electricity Networks: An Overview. *IEEE Transactions on Smart Grid*, *11*(4), 3185–3200. doi:10.1109/TSG.2020.2969657

Tushar, W., Yuen, C., Mohsenian-Rad, H., Saha, T., Poor, H. V., & Wood, K. L. (2018). Transforming energy networks via peer-to-peer energy trading: The potential of game-theoretic approaches. *IEEE Signal Processing Magazine*, *35*(4), 90–111. doi:10.1109/MSP.2018.2818327

UNDP. (2012). *Climate Public Expenditure and Institutional Reviews (CPEIRs) in the Asia-Pacific Region - What have We Learnt?* UNDP. https://www.asia-pacific.undp.org/content/dam/rbap/docs/Research%20&%20Publications/democratic_governance/APRC-DG-2012-CPEIRLessonsLearnt.pdf

UNDP. (2016a). *Charting New Territory: A stock take of climate change finance frameworks in Asia-Pacific*. UNDP. https://www.climatefinance-developmenteffectiveness.org/sites/default/files/documents/09_06_16/Charting%20New%20Territory%20%20A%20Stocktake%20of%20Climate%20Change%20Financing%20Frameworks%20in%20Asia%20Pacific.pdf

UNDP. (2016b). *The Adaptation Finance Gap Report*. UNDP. orbit.dtu.dk/ws/files/177810752/50313_UNEP_GAP_report_2016_v5_SB.pdf

UNECA. (2022). *Nationally Determined Contributions (NDCs)*. United Nations Economic Commission for Africa. www.uneca.org. https://www.uneca.org/african-climate-policy-centre/nationally-determined-contributions-%28ndcs%29

UNFCCC. (2018). *UNFCCC Standing Committee on Finance 2018 Biennial Assessment and Overview of Climate Finance Flows*. UNFCCC. https://unfccc.int/sites/default/files/resource/2018%20BA%20Technical%20Report%20Final.pdf

UNFPA. (2009). *Financing that Makes a Difference*. UNFPA. https://www.unfpa.org/sites/default/files/pub-pdf/climateconnections_5_finance.pdf

United Nations Economic Commission for Africa. (n.d.). *African small island developing states*. UNECA. https://archive.uneca.org/africansmallislanddevelopingstates/pages/african-small-island-developing-states

United Nations Framework Convention on Climate Change (UNFCCC). (2016). *The Paris Agreement*. UNFCC. https://unfccc.int/sites/default/files/resource/parisagreement_publication.pdf

United Nations. (2018). *According to UN experts, Indigenous rights must be respected during Kenya's climate change project*. OHCHR. https://www.ohchr.org/en/press-releases/2018/01/indigenous-rights-must-be-respected-during-kenya-climate-change-project-say

United Nations. (2023). *The Paris Agreement*. UN. https://unfccc.int/process-and-meetings/the-paris-agreement

UNOSSC. (2021). *Good Practices in South-South and Triangular Cooperation for Sustainable Development in SIDS*. UNOSSC. https://unctad.org/news/small-island-developing-states-face-uphill-battle-covid-19-recovery

Urwin, K., & Jordan, A. (2008). Does public policy support or undermine climate change adaptation? Exploring policy interplay across different scales of governance. *Global Environmental Change*, *18*(1), 180–191. doi:10.1016/j.gloenvcha.2007.08.002

US DoE. (2017). Annual Energy Outlook. United States Department of Energy.

USAID. (2020). *South Africa Power Africa Fact Sheet*. USAID. https://www.usaid.gov/powerafrica/south-africa

USAID. (2021). *Nigeria, Climate Change Country Profile*. USAID. https://www.usaid.gov/climate/country-profiles/nigeria#:~:text=Its%20multiple%20ecological%20zones%20have,to%20flooding%20and%20waterborne%20disease

van Soest, H. (2019). *Peer-to-peer electricity trading: A review of the legal context*. Competition and Regulation in Network Industries., doi:10.1177/1783591719834902

Vogel, B., & Henstra, D. (2015). Studying local climate adaptation: A heuristic research framework for comparative policy analysis. *Global Environmental Change*, *31*, 110–120. doi:10.1016/j.gloenvcha.2015.01.001

Wang, K., Yan, X., Yuan, Y., & Tang, D. (2018). Optimizing Ship Energy Efficiency: Application of Particle Swarm Optimization Algorithm. Proc. Institution Mech. Eng. Part M J. *Proceedings of the Institution of Mechanical Engineers. Proceedings Part M, Journal of Engineering for the Maritime Environment*, *232*(4), 379–391. doi:10.1177/1475090216638879

WAPP. (2023). *West African Power Pool*. WAPP. https://www.ecowapp.org/

Williams, M. J. (2017). The political economy of unfinished development projects: Corruption, clientelism, or collective choice? *The American Political Science Review*, *111*(4), 705–723. doi:10.1017/S0003055417000351

WIPO. (2020). *World Intellectual Property Organization (WIPO) Global Innovation Index 2020*. WIPO. https://www.wipo.int/edocs/pubdocs/en/wipo_pub_gii_2020/ng.pdf

Wongthongtham, P., Marrable, D., Abu-Salih, B., Liu, X., & Morrison, G. (2021). Blockchain-enabled Peer-to-Peer energy trading. *Computers and Electrical Engineering, 94*(September 2020), 107299. doi:10.1016/j.compeleceng.2021.107299

World Bank. (2010). *World Development Report 2010. Development and Climate Change*. World Bank.

World Bank. (2018). Off-grid Solar Market Trends Report 2018. World Bank. doi:10.1017/CBO9781107415324.004

World Bank. (2020). *Doing Business*. World Bank. https://documents1.worldbank.org/curated/en/688761571934946384/pdf/Doing Business 2020-Comparing-Business-Regulation-in-190-Economies.pdf

World Bank. (2021). *Nigeria to Improve Electricity Access and Services to Citizens*. World Bank. https://www.worldbank.org/en/news/press release/2021/02/05/Nigeria-to-improve-electricity-access-and-services-to citizens

World Bank. (2023). *Leveraging Natural Wealth to Build Opportunities*. World Bank. https://documents1.worldbank.org/curated/en/099457104212338173/pdf/IDU0d4e660690db44049cc0bd440699b9c1b4544.pdf

World Bank. (n.d.). *Sao Tome and Principe Overview: Development news, research, data*. World Bank. https://www.worldbank.org/en/country/saotome/overview

World Energy Council. (2018). *The Role of ICT in Energy Efficiency Management: Household Sector*. World Energy Council. https://www.worldenergy.org/assets/downloads/20180420_TF_paper_final.pdf

World Energy Resources. (2013). *Geothermal*. World Energy Council 2013. https://www.worldenergy.org/assets/images/imported/2013/10/WER_2013_9_Geothermal.pdf

WorldBank. (2022). *Sub-Saharan Africa* (pp. 4–17). Washington, D.C., United States: WorldBank. https://databank.worldbank.org/data/download/poverty/33EF03BB-9722-4AE2-ABC7-AA2972D68AFE/Global_POVEQ_SSA.pdf

Yang, Y., Ren, J., Stubbe, H., Xu, D. & Tien, T. (2018). Using multi-criteria analysis to prioritize renewable energy home heating technologies. CBS Research Portal.

Yong, C. (2022). *About Africa*. UNDP in Africa. https://web.archive.org/web/20200411014537/https://www.africa.undp.org/content/rba/en/home/regioninfo.html https://climatepromise.undp.org/news-and-stories/without-respecting-rights-and-knowledge-indigenous-peoples-climate-pledges-will

Zacharoula, S. A. (2012). Green Informatics: ICT for Green and Sustainability. *Agrárinformatika Folyóirat*, *3*(2). http://real.mtak.hu/23913/1/89_371_1_PB_u.pdf

Zeeshan, M., Han, J., Rehman, A., Ullah, I., & Mubashir, M. (2022). Exploring the Role of Information Communication Technology and Renewable Energy in Environmental Quality of South-East Asian Emerging Economies. *Frontiers in Environmental Science*, *10*, 917468. doi:10.3389/fenvs.2022.917468

Zhang, C., Wu, J., Zhou, Y., Cheng, M., & Long, C. (2018). Peer-to-Peer energy trading in a Microgrid. *Applied Energy*, *220*(June), 1–12. doi:10.1016/j.apenergy.2018.03.010

Zhou, Y., Wu, J., Long, C., & Ming, W. (2020). State-of-the-Art Analysis and Perspectives for Peer-to-Peer Energy Trading. *Engineering (Beijing)*, *6*(7), 739–753. doi:10.1016/j.eng.2020.06.002

Ziervogel, G., Johnston, P., Matthew, M., & Mukheibir, P. (2010). Using climate information for supporting climate change adaptation in water resource management in South Africa. *Climatic Change*, *103*(3-4), 537–554. doi:10.100710584-009-9771-3

Zyl, N. P. (2020, August 13). *Chad solar power projects earmarked to support N'Djamena surrounds in Chad*. ESI-Africa. https://www.esi-africa.com/industry-sectors/generation/solar/solar-projects-earmarked-to-support-ndjamena-surrounds-in-chad/

Related References

To continue our tradition of advancing academic research, we have compiled a list of recommended IGI Global readings. These references will provide additional information and guidance to further enrich your knowledge and assist you with your own research and future publications.

Ajiboye, O. E., & Yusuff, O. S. (2017). Foreign Land Acquisition: Food Security and Food Chains – The Nigerian Experience. In I. Management Association (Ed.), Natural Resources Management: Concepts, Methodologies, Tools, and Applications (pp. 1524-1545). Hershey, PA: IGI Global. https://doi.org/ doi:10.4018/978-1-5225-0803-8.ch072

Alapiki, H. E., & Amadi, L. A. (2018). Sustainable Food Consumption in the Neoliberal Order: Challenges and Policy Implications. In A. Obayelu (Ed.), *Food Systems Sustainability and Environmental Policies in Modern Economies* (pp. 90–123). Hershey, PA: IGI Global. doi:10.4018/978-1-5225-3631-4.ch005

Altaş, A. (2018). Geographical Information System Applications Utilized in Museums in Turkey Within the Scope of the Cultural Heritage Tourism: A Case Study of Mobile Application of Müze Asist. In S. Chaudhuri & N. Ray (Eds.), *GIS Applications in the Tourism and Hospitality Industry* (pp. 42–60). Hershey, PA: IGI Global. doi:10.4018/978-1-5225-5088-4.ch002

Andreea, I. R. (2018). Beyond Macroeconomics of Food and Nutrition Security. *International Journal of Sustainable Economies Management*, *7*(1), 13–22. doi:10.4018/IJSEM.2018010102

Anwar, J. (2017). Reproductive and Mental Health during Natural Disaster: Implications and Issues for Women in Developing Nations – A Case Example. In I. Management Association (Ed.), Gaming and Technology Addiction: Breakthroughs in Research and Practice (pp. 446-472). Hershey, PA: IGI Global. https://doi.org/ doi:10.4018/978-1-5225-0778-9.ch021

Awadh, H., Aksissou, M., Benhardouze, W., Darasi, F., & Snaiki, J. (2018). Socioeconomic Status of Artisanal Fishers in the West Part of Moroccan Mediterranean. *International Journal of Social Ecology and Sustainable Development*, *9*(1), 40–52. doi:10.4018/IJSESD.2018010104

Aye, G. C., & Haruna, R. F. (2018). Effect of Climate Change on Crop Productivity and Prices in Benue State, Nigeria: Implications for Food Security. In V. Erokhin (Ed.), *Establishing Food Security and Alternatives to International Trade in Emerging Economies* (pp. 244–268). Hershey, PA: IGI Global. doi:10.4018/978-1-5225-2733-6.ch012

Azizan, S. A., & Suki, N. M. (2017). Consumers' Intentions to Purchase Organic Food Products. In T. Esakki (Ed.), *Green Marketing and Environmental Responsibility in Modern Corporations* (pp. 68–82). Hershey, PA: IGI Global. doi:10.4018/978-1-5225-2331-4.ch005

Barakabitze, A. A., Fue, K. G., Kitindi, E. J., & Sanga, C. A. (2017). Developing a Framework for Next Generation Integrated Agro Food-Advisory Systems in Developing Countries. In I. Management Association (Ed.), Agri-Food Supply Chain Management: Breakthroughs in Research and Practice (pp. 47-67). Hershey, PA: IGI Global. doi:10.4018/978-1-5225-1629-3.ch004

Beachcroft-Shaw, H., & Ellis, D. (2018). Using Successful Cases to Promote Environmental Sustainability: A Social Marketing Approach. In I. Management Association (Ed.), Sustainable Development: Concepts, Methodologies, Tools, and Applications (pp. 936-953). Hershey, PA: IGI Global. doi:10.4018/978-1-5225-3817-2.ch042

Behnassi, M., Kahime, K., Boussaa, S., Boumezzough, A., & Messouli, M. (2017). Infectious Diseases and Climate Vulnerability in Morocco: Governance and Adaptation Options. In I. Management Association (Ed.), Public Health and Welfare: Concepts, Methodologies, Tools, and Applications (pp. 91-109). Hershey, PA: IGI Global. https://doi.org/ doi:10.4018/978-1-5225-1674-3.ch005

Bekele, F., & Bekele, I. (2017). Social and Environmental Impacts on Agricultural Development. In W. Ganpat, R. Dyer, & W. Isaac (Eds.), *Agricultural Development and Food Security in Developing Nations* (pp. 21–56). Hershey, PA: IGI Global. doi:10.4018/978-1-5225-0942-4.ch002

Benaouda, A., & García-Peñalvo, F. J. (2018). Towards an Intelligent System for the Territorial Planning: Agricultural Case. In F. García-Peñalvo (Ed.), *Global Implications of Emerging Technology Trends* (pp. 158–178). Hershey, PA: IGI Global. doi:10.4018/978-1-5225-4944-4.ch010

Bhaskar, A., Rao, G. B., & Vencatesan, J. (2017). Characterization and Management Concerns of Water Resources around Pallikaranai Marsh, South Chennai. In P. Rao & Y. Patil (Eds.), *Reconsidering the Impact of Climate Change on Global Water Supply, Use, and Management* (pp. 102–121). Hershey, PA: IGI Global. doi:10.4018/978-1-5225-1046-8.ch007

Bhyan, P., Shrivastava, B., & Kumar, N. (2022). Requisite Sustainable Development Contemplating Buildings: Economic and Environmental Sustainability. In A. Hussain, K. Tiwari, & A. Gupta (Eds.), *Addressing Environmental Challenges Through Spatial Planning* (pp. 269–288). IGI Global. https://doi.org/10.4018/978-1-7998-8331-9.ch014

Bogataj, D., & Drobne, D. (2017). Control of Perishable Goods in Cold Logistic Chains by Bionanosensors. In S. Joo (Ed.), *Applying Nanotechnology for Environmental Sustainability* (pp. 376–402). Hershey, PA: IGI Global. doi:10.4018/978-1-5225-0585-3.ch016

Bogataj, D., & Drobne, D. (2017). Control of Perishable Goods in Cold Logistic Chains by Bionanosensors. In I. Management Association (Ed.), Materials Science and Engineering: Concepts, Methodologies, Tools, and Applications (pp. 471-497). Hershey, PA: IGI Global. https://doi.org/ doi:10.4018/978-1-5225-1798-6.ch019

Bogueva, D., & Marinova, D. (2018). What Is More Important: Perception of Masculinity or Personal Health and the Environment? In D. Bogueva, D. Marinova, & T. Raphaely (Eds.), *Handbook of Research on Social Marketing and Its Influence on Animal Origin Food Product Consumption* (pp. 148–162). Hershey, PA: IGI Global. doi:10.4018/978-1-5225-4757-0.ch010

Bouzid, M. (2017). Waterborne Diseases and Climate Change: Impact and Implications. In M. Bouzid (Ed.), *Examining the Role of Environmental Change on Emerging Infectious Diseases and Pandemics* (pp. 89–108). Hershey, PA: IGI Global. doi:10.4018/978-1-5225-0553-2.ch004

Bowles, D. C. (2017). Climate Change-Associated Conflict and Infectious Disease. In M. Bouzid (Ed.), *Examining the Role of Environmental Change on Emerging Infectious Diseases and Pandemics* (pp. 68–88). Hershey, PA: IGI Global. doi:10.4018/978-1-5225-0553-2.ch003

Buck, J. J., & Lowry, R. K. (2017). Oceanographic Data Management: Quills and Free Text to the Digital Age and "Big Data". In P. Diviacco, A. Leadbetter, & H. Glaves (Eds.), *Oceanographic and Marine Cross-Domain Data Management for Sustainable Development* (pp. 1–22). Hershey, PA: IGI Global. doi:10.4018/978-1-5225-0700-0.ch001

Buse, C. G. (2017). Are Climate Change Adaptation Policies a Game Changer?: A Case Study of Perspectives from Public Health Officials in Ontario, Canada. In M. Bouzid (Ed.), *Examining the Role of Environmental Change on Emerging Infectious Diseases and Pandemics* (pp. 230–257). Hershey, PA: IGI Global. doi:10.4018/978-1-5225-0553-2.ch010

Calderon, F. A., Giolo, E. G., Frau, C. D., Rengel, M. G., Rodriguez, H., Tornello, M., ... Gallucci, R. (2018). Seismic Microzonation and Site Effects Detection Through Microtremors Measures: A Review. In N. Ceryan (Ed.), *Handbook of Research on Trends and Digital Advances in Engineering Geology* (pp. 326–349). Hershey, PA: IGI Global. doi:10.4018/978-1-5225-2709-1.ch009

Carfì, D., Donato, A., & Panuccio, D. (2018). A Game Theory Coopetitive Perspective for Sustainability of Global Feeding: Agreements Among Vegan and Non-Vegan Food Firms. In I. Management Association (Ed.), Game Theory: Breakthroughs in Research and Practice (pp. 71-104). Hershey, PA: IGI Global. https://doi.org/doi:10.4018/978-1-5225-2594-3.ch004

Castagnolo, V. (2018). Analyzing, Classifying, Safeguarding: Drawing for the Borgo Murattiano Neighbourhood of Bari. In G. Carlone, N. Martinelli, & F. Rotondo (Eds.), *Designing Grid Cities for Optimized Urban Development and Planning* (pp. 93–108). Hershey, PA: IGI Global. doi:10.4018/978-1-5225-3613-0.ch006

Chekima, B. (2018). The Dilemma of Purchase Intention: A Conceptual Framework for Understanding Actual Consumption of Organic Food. *International Journal of Sustainable Economies Management*, *7*(2), 1–13. doi:10.4018/IJSEM.2018040101

Chen, Y. (2017). Sustainable Supply Chains and International Soft Landings: A Case of Wetland Entrepreneurship. In B. Christiansen & F. Kasarcı (Eds.), *Corporate Espionage, Geopolitics, and Diplomacy Issues in International Business* (pp. 232–247). Hershey, PA: IGI Global. doi:10.4018/978-1-5225-1031-4.ch013

Çıtak, L., Akel, V., & Ersoy, E. (2018). Investors' Reactions to the Announcement of New Constituents of BIST Sustainability Index: An Analysis by Event Study and Mean-Median Tests. In M. Risso & S. Testarmata (Eds.), *Value Sharing for Sustainable and Inclusive Development* (pp. 270–289). Hershey, PA: IGI Global. doi:10.4018/978-1-5225-3147-0.ch012

D'Aleo, V., D'Aleo, F., & Bonanno, R. (2018). New Food Industries Toward a New Level of Sustainable Supply: Success Stories, Business Models, and Strategies. In V. Erokhin (Ed.), *Establishing Food Security and Alternatives to International Trade in Emerging Economies* (pp. 74–97). Hershey, PA: IGI Global. doi:10.4018/978-1-5225-2733-6.ch004

Dagevos, H., & Reinders, M. J. (2018). Flexitarianism and Social Marketing: Reflections on Eating Meat in Moderation. In D. Bogueva, D. Marinova, & T. Raphaely (Eds.), *Handbook of Research on Social Marketing and Its Influence on Animal Origin Food Product Consumption* (pp. 105–120). Hershey, PA: IGI Global. doi:10.4018/978-1-5225-4757-0.ch007

Danisman, G. O. (2022). What Drives Eco-Design Innovations in European SMEs? In U. Akkucuk (Ed.), *Disruptive Technologies and Eco-Innovation for Sustainable Development* (pp. 191–206). IGI Global. https://doi.org/10.4018/978-1-7998-8900-7.ch011

Deenapanray, P. N., & Ramma, I. (2017). Adaptations to Climate Change and Climate Variability in the Agriculture Sector in Mauritius: Lessons from a Technical Needs Assessment. In I. Management Association (Ed.), Natural Resources Management: Concepts, Methodologies, Tools, and Applications (pp. 655-680). Hershey, PA: IGI Global. https://doi.org/ doi:10.4018/978-1-5225-0803-8.ch030

Deenapanray, P. N., & Ramma, I. (2017). Adaptations to Climate Change and Climate Variability in the Agriculture Sector in Mauritius: Lessons from a Technical Needs Assessment. In I. Management Association (Ed.), Natural Resources Management: Concepts, Methodologies, Tools, and Applications (pp. 655-680). Hershey, PA: IGI Global. https://doi.org/ doi:10.4018/978-1-5225-0803-8.ch030

Deshpande, S., Basu, S. K., Li, X., & Chen, X. (2017). Smart, Innovative and Intelligent Technologies Used in Drug Designing. In I. Management Association (Ed.), Pharmaceutical Sciences: Breakthroughs in Research and Practice (pp. 1175-1191). Hershey, PA: IGI Global. https://doi.org/ doi:10.4018/978-1-5225-1762-7.ch045

Dlamini, P. N. (2017). Use of Information and Communication Technologies Tools to Capture, Store, and Disseminate Indigenous Knowledge: A Literature Review. In P. Ngulube (Ed.), *Handbook of Research on Theoretical Perspectives on Indigenous Knowledge Systems in Developing Countries* (pp. 225–247). Hershey, PA: IGI Global. doi:10.4018/978-1-5225-0833-5.ch010

Dolejsova, M., & Kera, D. (2017). The Fermentation GutHub Project and the Internet of Microbes. In S. Konomi & G. Roussos (Eds.), *Enriching Urban Spaces with Ambient Computing, the Internet of Things, and Smart City Design* (pp. 25–46). Hershey, PA: IGI Global. doi:10.4018/978-1-5225-0827-4.ch002

Dolunay, O. (2018). A Paradigm Shift: Empowering Farmers to Eliminate the Waste in the Form of Fresh Water and Energy Through the Implementation of 4R+T. In I. Management Association (Ed.), Sustainable Development: Concepts, Methodologies, Tools, and Applications (pp. 882-892). Hershey, PA: IGI Global. https://doi.org/ doi:10.4018/978-1-5225-3817-2.ch039

Dube, P., Heijman, W. J., Ihle, R., & Ochieng, J. (2018). The Potential of Traditional Leafy Vegetables for Improving Food Security in Africa. In V. Erokhin (Ed.), *Establishing Food Security and Alternatives to International Trade in Emerging Economies* (pp. 220–243). Hershey, PA: IGI Global. doi:10.4018/978-1-5225-2733-6.ch011

Duruji, M. M., & Urenma, D. F. (2017). The Environmentalism and Politics of Climate Change: A Study of the Process of Global Convergence through UNFCCC Conferences. In I. Management Association (Ed.), *Natural Resources Management: Concepts, Methodologies, Tools, and Applications* (pp. 77-108). Hershey, PA: IGI Global. 10.4018/978-1-5225-0803-8.ch004

Dutta, U. (2017). Agro-Geoinformatics, Potato Cultivation, and Climate Change. In S. Londhe (Ed.), *Sustainable Potato Production and the Impact of Climate Change* (pp. 247–271). Hershey, PA: IGI Global. doi:10.4018/978-1-5225-1715-3.ch012

Edirisinghe, R., Stranieri, A., & Wickramasinghe, N. (2017). A Taxonomy for mHealth. In N. Wickramasinghe (Ed.), *Handbook of Research on Healthcare Administration and Management* (pp. 596–615). Hershey, PA: IGI Global. doi:10.4018/978-1-5225-0920-2.ch036

Ekpeni, N. M., & Ayeni, A. O. (2018). Global Natural Hazard and Disaster Vulnerability Management. In A. Eneanya (Ed.), *Handbook of Research on Environmental Policies for Emergency Management and Public Safety* (pp. 83–104). Hershey, PA: IGI Global. doi:10.4018/978-1-5225-3194-4.ch005

Ene, C., Voica, M. C., & Panait, M. (2017). Green Investments and Food Security: Opportunities and Future Directions in the Context of Sustainable Development. In M. Mieila (Ed.), *Measuring Sustainable Development and Green Investments in Contemporary Economies* (pp. 163–200). Hershey, PA: IGI Global. doi:10.4018/978-1-5225-2081-8.ch007

Escamilla, I., Ruíz, M. T., Ibarra, M. M., Soto, V. L., Quintero, R., & Guzmán, G. (2018). Geocoding Tweets Based on Semantic Web and Ontologies. In M. Lytras, N. Aljohani, E. Damiani, & K. Chui (Eds.), *Innovations, Developments, and Applications of Semantic Web and Information Systems* (pp. 372–392). Hershey, PA: IGI Global. doi:10.4018/978-1-5225-5042-6.ch014

Escribano, A. J. (2018). Marketing Strategies for Trendy Animal Products: Sustainability as a Core. In F. Quoquab, R. Thurasamy, & J. Mohammad (Eds.), *Driving Green Consumerism Through Strategic Sustainability Marketing* (pp. 169–203). Hershey, PA: IGI Global. doi:10.4018/978-1-5225-2912-5.ch010

Eudoxie, G., & Roopnarine, R. (2017). Climate Change Adaptation and Disaster Risk Management in the Caribbean. In W. Ganpat & W. Isaac (Eds.), *Environmental Sustainability and Climate Change Adaptation Strategies* (pp. 97–125). Hershey, PA: IGI Global. doi:10.4018/978-1-5225-1607-1.ch004

Farmer, L. S. (2017). Data Analytics for Strategic Management: Getting the Right Data. In V. Wang (Ed.), *Encyclopedia of Strategic Leadership and Management* (pp. 810–822). Hershey, PA: IGI Global. doi:10.4018/978-1-5225-1049-9.ch056

Fattal, L. R. (2017). Catastrophe: An Uncanny Catalyst for Creativity. In R. Shin (Ed.), *Convergence of Contemporary Art, Visual Culture, and Global Civic Engagement* (pp. 244–262). Hershey, PA: IGI Global. doi:10.4018/978-1-5225-1665-1.ch014

Forti, I. (2017). A Cross Reading of Landscape through Digital Landscape Models: The Case of Southern Garda. In A. Ippolito (Ed.), *Handbook of Research on Emerging Technologies for Architectural and Archaeological Heritage* (pp. 532–561). Hershey, PA: IGI Global. doi:10.4018/978-1-5225-0675-1.ch018

Gharbi, A., De Runz, C., & Akdag, H. (2017). Urban Development Modelling: A Survey. In S. Faiz & K. Mahmoudi (Eds.), *Handbook of Research on Geographic Information Systems Applications and Advancements* (pp. 96–124). Hershey, PA: IGI Global. doi:10.4018/978-1-5225-0937-0.ch004

Ghosh, I., & Ghoshal, I. (2018). Implications of Trade Liberalization for Food Security Under the ASEAN-India Strategic Partnership: A Gravity Model Approach. In V. Erokhin (Ed.), *Establishing Food Security and Alternatives to International Trade in Emerging Economies* (pp. 98–118). Hershey, PA: IGI Global. doi:10.4018/978-1-5225-2733-6.ch005

Glaves, H. M. (2017). Developing a Common Global Framework for Marine Data Management. In P. Diviacco, A. Leadbetter, & H. Glaves (Eds.), *Oceanographic and Marine Cross-Domain Data Management for Sustainable Development* (pp. 47–68). Hershey, PA: IGI Global. doi:10.4018/978-1-5225-0700-0.ch003

Godulla, A., & Wolf, C. (2018). Future of Food: Transmedia Strategies of National Geographic. In R. Gambarato & G. Alzamora (Eds.), *Exploring Transmedia Journalism in the Digital Age* (pp. 162–182). Hershey, PA: IGI Global. doi:10.4018/978-1-5225-3781-6.ch010

Gomes, P. P. (2018). Food and Environment: A Review on the Sustainability of Six Different Dietary Patterns. In A. Obayelu (Ed.), *Food Systems Sustainability and Environmental Policies in Modern Economies* (pp. 15–31). Hershey, PA: IGI Global. doi:10.4018/978-1-5225-3631-4.ch002

Gonzalez-Feliu, J. (2018). Sustainability Evaluation of Green Urban Logistics Systems: Literature Overview and Proposed Framework. In A. Paul, D. Bhattacharyya, & S. Anand (Eds.), *Green Initiatives for Business Sustainability and Value Creation* (pp. 103–134). Hershey, PA: IGI Global. doi:10.4018/978-1-5225-2662-9.ch005

Goodland, R. (2017). A Fresh Look at Livestock Greenhouse Gas Emissions and Mitigation. In I. Management Association (Ed.), Natural Resources Management: Concepts, Methodologies, Tools, and Applications (pp. 124-139). Hershey, PA: IGI Global. doi:10.4018/978-1-5225-0803-8.ch006

Goundar, S., & Appana, S. (2018). Mainstreaming Development Policies for Climate Change in Fiji: A Policy Gap Analysis and the Role of ICTs. In I. Management Association (Ed.), Sustainable Development: Concepts, Methodologies, Tools, and Applications (pp. 402-432). Hershey, PA: IGI Global. doi:10.4018/978-1-5225-3817-2.ch020

Granell-Canut, C., & Aguilar-Moreno, E. (2018). Geospatial Influence in Science Mapping. In M. Khosrow-Pour, D.B.A. (Ed.), Encyclopedia of Information Science and Technology, Fourth Edition (pp. 3473-3483). Hershey, PA: IGI Global. doi:10.4018/978-1-5225-2255-3.ch302

Grigelis, A., Blažauskas, N., Gelumbauskaitė, L. Ž., Gulbinskas, S., Suzdalev, S., & Ferrarin, C. (2017). Marine Environment Data Management Related to the Human Activity in the South-Eastern Baltic Sea (The Lithuanian Sector). In P. Diviacco, A. Leadbetter, & H. Glaves (Eds.), *Oceanographic and Marine Cross-Domain Data Management for Sustainable Development* (pp. 282–302). Hershey, PA: IGI Global. doi:10.4018/978-1-5225-0700-0.ch012

Guma, I. P., Rwashana, A. S., & Oyo, B. (2018). Food Security Policy Analysis Using System Dynamics: The Case of Uganda. *International Journal of Information Technologies and Systems Approach*, *11*(1), 72–90. doi:10.4018/IJITSA.2018010104

Guma, I. P., Rwashana, A. S., & Oyo, B. (2018). Food Security Indicators for Subsistence Farmers Sustainability: A System Dynamics Approach. *International Journal of System Dynamics Applications*, *7*(1), 45–64. doi:10.4018/IJSDA.2018010103

Gupta, P., & Goyal, S. (2017). Wildlife Habitat Evaluation. In A. Santra & S. Mitra (Eds.), *Remote Sensing Techniques and GIS Applications in Earth and Environmental Studies* (pp. 258–264). Hershey, PA: IGI Global. doi:10.4018/978-1-5225-1814-3.ch013

Hanson, T., & Hildebrand, E. (2018). GPS Travel Diaries in Rural Transportation Research: A Focus on Older Drivers. In I. Management Association (Ed.), Intelligent Transportation and Planning: Breakthroughs in Research and Practice (pp. 609-625). Hershey, PA: IGI Global. doi:10.4018/978-1-5225-5210-9.ch027

Hartman, M. B. (2017). Research-Based Climate Change Public Education Programs. In I. Management Association (Ed.), Natural Resources Management: Concepts, Methodologies, Tools, and Applications (pp. 992-1003). Hershey, PA: IGI Global. doi:10.4018/978-1-5225-0803-8.ch046

Hashim, N. (2017). Zanzibari Seaweed: Global Climate Change and the Promise of Adaptation. In I. Management Association (Ed.), Natural Resources Management: Concepts, Methodologies, Tools, and Applications (pp. 365-391). Hershey, PA: IGI Global. https://doi.org/ doi:10.4018/978-1-5225-0803-8.ch019

Herrera, J. E., Argüello, L. V., Gonzalez-Feliu, J., & Jaimes, W. A. (2017). Decision Support System Design Requirements, Information Management, and Urban Logistics Efficiency: Case Study of Bogotá, Colombia. In G. Jamil, A. Soares, & C. Pessoa (Eds.), *Handbook of Research on Information Management for Effective Logistics and Supply Chains* (pp. 223–238). Hershey, PA: IGI Global. doi:10.4018/978-1-5225-0973-8.ch012

Huizinga, T., Ayanso, A., Smoor, M., & Wronski, T. (2017). Exploring Insurance and Natural Disaster Tweets Using Text Analytics. *International Journal of Business Analytics*, *4*(1), 1–17. doi:10.4018/IJBAN.2017010101

Hung, K., Kalantari, M., & Rajabifard, A. (2017). An Integrated Method for Assessing the Text Content Quality of Volunteered Geographic Information in Disaster Management. *International Journal of Information Systems for Crisis Response and Management*, *9*(2), 1–17. doi:10.4018/IJISCRAM.2017040101

Husnain, A., & Avdic, A. (2018). Identifying the Contemporary Status of E-Service Sustainability Research. In I. Management Association (Ed.), Sustainable Development: Concepts, Methodologies, Tools, and Applications (pp. 467-485). Hershey, PA: IGI Global. https://doi.org/ doi:10.4018/978-1-5225-3817-2.ch022

Iarossi, M. P., & Ferro, L. (2017). "The Past is Never Dead. It's Not Even Past": Virtual Archaeological Promenade. In A. Ippolito & M. Cigola (Eds.), *Handbook of Research on Emerging Technologies for Digital Preservation and Information Modeling* (pp. 228–255). Hershey, PA: IGI Global. doi:10.4018/978-1-5225-0680-5.ch010

Ignjatijević, S., & Cvijanović, D. (2018). Analysis of Serbian Honey Production and Exports. In *Exploring the Global Competitiveness of Agri-Food Sectors and Serbia's Dominant Presence: Emerging Research and Opportunities* (pp. 109–139). Hershey, PA: IGI Global. doi:10.4018/978-1-5225-2762-6.ch005

Jana, S. K., & Karmakar, A. K. (2017). Globalization, Governance, and Food Security: The Case of BRICS. In I. Management Association (Ed.), Natural Resources Management: Concepts, Methodologies, Tools, and Applications (pp. 692-712). Hershey, PA: IGI Global. https://doi.org/ doi:10.4018/978-1-5225-0803-8.ch032

Jana, S. K., & Karmakar, A. K. (2017). Food Security in Asia: Is There Convergence? In I. Management Association (Ed.), Natural Resources Management: Concepts, Methodologies, Tools, and Applications (pp. 109-123). Hershey, PA: IGI Global. https://doi.org/ doi:10.4018/978-1-5225-0803-8.ch005

John, J., & Kumar, S. (2018). A Locational Decision Making Framework for Shipbreaking Under Multiple Criteria. In I. Management Association (Ed.), Operations and Service Management: Concepts, Methodologies, Tools, and Applications (pp. 504-527). Hershey, PA: IGI Global. doi:10.4018/978-1-5225-3909-4.ch024

John, J., & Srivastava, R. K. (2018). Decision Insights for Shipbreaking using Environmental Impact Assessment: Review and Perspectives. *International Journal of Strategic Decision Sciences*, *9*(1), 45–62. doi:10.4018/IJSDS.2018010104

Joshi, Y., & Rahman, Z. (2018). Determinants of Sustainable Consumption Behaviour: Review and Conceptual Framework. In A. Paul, D. Bhattacharyya, & S. Anand (Eds.), *Green Initiatives for Business Sustainability and Value Creation* (pp. 239–262). Hershey, PA: IGI Global. doi:10.4018/978-1-5225-2662-9.ch011

Juma, D. W., Reuben, M., Wang, H., & Li, F. (2018). Adaptive Coevolution: Realigning the Water Governance Regime to the Changing Climate. In I. Management Association (Ed.), Hydrology and Water Resource Management: Breakthroughs in Research and Practice (pp. 346-357). Hershey, PA: IGI Global. https://doi.org/ doi:10.4018/978-1-5225-3427-3.ch014

K., S., & Tripathy, B. K. (2018). Neighborhood Rough-Sets-Based Spatial Data Analytics. In M. Khosrow-Pour, D.B.A. (Ed.), *Encyclopedia of Information Science and Technology, Fourth Edition* (pp. 1835-1844). Hershey, PA: IGI Global. https:// doi.org/ doi:10.4018/978-1-5225-2255-3.ch160

Kabir, F. (2018). Towards a More Gender-Inclusive Climate Change Policy. In N. Mahtab, T. Haque, I. Khan, M. Islam, & I. Wahid (Eds.), *Handbook of Research on Women's Issues and Rights in the Developing World* (pp. 354–369). Hershey, PA: IGI Global. doi:10.4018/978-1-5225-3018-3.ch022

Kabir, F. (2018). Towards a More Gender-Inclusive Climate Change Policy. In I. Management Association (Ed.), Climate Change and Environmental Concerns: Breakthroughs in Research and Practice (pp. 525-540). Hershey, PA: IGI Global. https://doi.org/ doi:10.4018/978-1-5225-5487-5.ch027

Kanyamuka, J. S., Jumbe, C. B., & Ricker-Gilbert, J. (2018). Making Agricultural Input Subsidies More Effective and Profitable in Africa: The Role of Complementary Interventions. In A. Obayelu (Ed.), *Food Systems Sustainability and Environmental Policies in Modern Economies* (pp. 172–187). Hershey, PA: IGI Global. doi:10.4018/978-1-5225-3631-4.ch008

Karimi, H., & Gholamrezafahimi, F. (2017). Study of Integrated Coastal Zone Management and Its Environmental Effects: A Case of Iran. In R. Singh, A. Singh, & V. Srivastava (Eds.), *Environmental Issues Surrounding Human Overpopulation* (pp. 64–88). Hershey, PA: IGI Global. doi:10.4018/978-1-5225-1683-5.ch004

Kaya, I. R., Hutabarat, J., & Bambang, A. N. (2018). "Sasi": A New Path to Sustain Seaweed Farming From Up-Stream to Down-Stream in Kotania Bay, Molucass. *International Journal of Social Ecology and Sustainable Development*, *9*(2), 28–36. doi:10.4018/IJSESD.2018040103

Khader, V. (2018). Technologies for Food, Health, Livelihood, and Nutrition Security. In I. Management Association (Ed.), Food Science and Nutrition: Breakthroughs in Research and Practice (pp. 94-112). Hershey, PA: IGI Global. https://doi.org/doi:10.4018/978-1-5225-5207-9.ch005

Kocadağlı, A. Y. (2017). The Temporal and Spatial Development of Organic Agriculture in Turkey. In W. Ganpat, R. Dyer, & W. Isaac (Eds.), *Agricultural Development and Food Security in Developing Nations* (pp. 130–156). Hershey, PA: IGI Global. doi:10.4018/978-1-5225-0942-4.ch006

Koundouri, P., Giannouli, A., & Souliotis, I. (2017). An Integrated Approach for Sustainable Environmental and Socio-Economic Development Using Offshore Infrastructure. In I. Management Association (Ed.), Renewable and Alternative Energy: Concepts, Methodologies, Tools, and Applications (pp. 1581-1601). Hershey, PA: IGI Global. doi:10.4018/978-1-5225-1671-2.ch056

Kumar, A., & Dash, M. K. (2017). Sustainability and Future Generation Infrastructure on Digital Platform: A Study of Generation Y. In N. Ray (Ed.), *Business Infrastructure for Sustainability in Developing Economies* (pp. 124–142). Hershey, PA: IGI Global. doi:10.4018/978-1-5225-2041-2.ch007

Kumar, A., Mukherjee, A. B., & Krishna, A. P. (2017). Application of Conventional Data Mining Techniques and Web Mining to Aid Disaster Management. In A. Kumar (Ed.), *Web Usage Mining Techniques and Applications Across Industries* (pp. 138–167). Hershey, PA: IGI Global. doi:10.4018/978-1-5225-0613-3.ch006

Kumar, C. P. (2017). Impact of Climate Change on Groundwater Resources. In I. Management Association (Ed.), Natural Resources Management: Concepts, Methodologies, Tools, and Applications (pp. 1094-1120). Hershey, PA: IGI Global. doi:10.4018/978-1-5225-0803-8.ch052

Kumari, S., & Patil, Y. (2017). Achieving Climate Smart Agriculture with a Sustainable Use of Water: A Conceptual Framework for Sustaining the Use of Water for Agriculture in the Era of Climate Change. In P. Rao & Y. Patil (Eds.), *Reconsidering the Impact of Climate Change on Global Water Supply, Use, and Management* (pp. 122–143). Hershey, PA: IGI Global. doi:10.4018/978-1-5225-1046-8.ch008

Kumari, S., & Patil, Y. (2018). Achieving Climate Smart Agriculture With a Sustainable Use of Water: A Conceptual Framework for Sustaining the Use of Water for Agriculture in the Era of Climate Change. In I. Management Association (Ed.), Climate Change and Environmental Concerns: Breakthroughs in Research and Practice (pp. 111-133). Hershey, PA: IGI Global. doi:10.4018/978-1-5225-5487-5.ch006

Kursah, M. B. (2017). Least-Cost Pipeline using Geographic Information System: The Limit to Technicalities. *International Journal of Applied Geospatial Research, 8*(3), 1–15. doi:10.4018/ijagr.2017070101

Lahiri, S., Ghosh, D., & Bhakta, J. N. (2017). Role of Microbes in Eco-Remediation of Perturbed Aquatic Ecosystem. In J. Bhakta (Ed.), *Handbook of Research on Inventive Bioremediation Techniques* (pp. 70–107). Hershey, PA: IGI Global. doi:10.4018/978-1-5225-2325-3.ch004

Lallo, C. H., Smalling, S., Facey, A., & Hughes, M. (2018). The Impact of Climate Change on Small Ruminant Performance in Caribbean Communities. In I. Management Association (Ed.), Climate Change and Environmental Concerns: Breakthroughs in Research and Practice (pp. 193-218). Hershey, PA: IGI Global. doi:10.4018/978-1-5225-5487-5.ch010

Laurini, R. (2017). Nature of Geographic Knowledge Bases. In S. Faiz & K. Mahmoudi (Eds.), *Handbook of Research on Geographic Information Systems Applications and Advancements* (pp. 29–60). Hershey, PA: IGI Global. doi:10.4018/978-1-5225-0937-0.ch002

Lawrence, J., Simpson, L., & Piggott, A. (2017). Protected Agriculture: A Climate Change Adaptation for Food and Nutrition Security. In I. Management Association (Ed.), Natural Resources Management: Concepts, Methodologies, Tools, and Applications (pp. 140-158). Hershey, PA: IGI Global. doi:10.4018/978-1-5225-0803-8.ch007

Leadbetter, A., Cheatham, M., Shepherd, A., & Thomas, R. (2017). Linked Ocean Data 2.0. In P. Diviacco, A. Leadbetter, & H. Glaves (Eds.), *Oceanographic and Marine Cross-Domain Data Management for Sustainable Development* (pp. 69–99). Hershey, PA: IGI Global. doi:10.4018/978-1-5225-0700-0.ch004

Lucas, M. R., Rego, C., Vieira, C., & Vieira, I. (2017). Proximity and Cooperation for Innovative Regional Development: The Case of the Science and Technology Park of Alentejo. In L. Carvalho (Ed.), *Handbook of Research on Entrepreneurial Development and Innovation Within Smart Cities* (pp. 199–228). Hershey, PA: IGI Global. doi:10.4018/978-1-5225-1978-2.ch010

Ma, X., Beaulieu, S. E., Fu, L., Fox, P., Di Stefano, M., & West, P. (2017). Documenting Provenance for Reproducible Marine Ecosystem Assessment in Open Science. In P. Diviacco, A. Leadbetter, & H. Glaves (Eds.), *Oceanographic and Marine Cross-Domain Data Management for Sustainable Development* (pp. 100–126). Hershey, PA: IGI Global. doi:10.4018/978-1-5225-0700-0.ch005

Mabe, L. K., & Oladele, O. I. (2017). Application of Information Communication Technologies for Agricultural Development through Extension Services: A Review. In T. Tossy (Ed.), *Information Technology Integration for Socio-Economic Development* (pp. 52–101). Hershey, PA: IGI Global. doi:10.4018/978-1-5225-0539-6.ch003

Malomo, B. I. (2018). A Review of Psychological Resilience as a Response to Natural Hazards in Nigeria. In A. Eneanya (Ed.), *Handbook of Research on Environmental Policies for Emergency Management and Public Safety* (pp. 147–165). Hershey, PA: IGI Global. doi:10.4018/978-1-5225-3194-4.ch008

Manchiraju, S. (2018). Predicting Behavioral Intentions Toward Sustainable Fashion Consumption: A Comparison of Attitude-Behavior and Value-Behavior Consistency Models. In I. Management Association (Ed.), Fashion and Textiles: Breakthroughs in Research and Practice (pp. 1-21). Hershey, PA: IGI Global. doi:10.4018/978-1-5225-3432-7.ch001

Manzella, G. M., Bartolini, R., Bustaffa, F., D'Angelo, P., De Mattei, M., Frontini, F., ... Spada, A. (2017). Semantic Search Engine for Data Management and Sustainable Development: Marine Planning Service Platform. In P. Diviacco, A. Leadbetter, & H. Glaves (Eds.), *Oceanographic and Marine Cross-Domain Data Management for Sustainable Development* (pp. 127–154). Hershey, PA: IGI Global. doi:10.4018/978-1-5225-0700-0.ch006

Mbonigaba, J. (2018). Comparing the Effects of Unsustainable Production and Consumption of Food on Health and Policy Across Developed and Less Developed Countries. In A. Obayelu (Ed.), *Food Systems Sustainability and Environmental Policies in Modern Economies* (pp. 124–158). Hershey, PA: IGI Global. doi:10.4018/978-1-5225-3631-4.ch006

McKeown, A. E. (2017). Nurses, Healthcare, and Environmental Pollution and Solutions: Breaking the Cycle of Harm. In I. Management Association (Ed.), Natural Resources Management: Concepts, Methodologies, Tools, and Applications (pp. 392-415). Hershey, PA: IGI Global. https://doi.org/ doi:10.4018/978-1-5225-0803-8.ch020

Mili, B., Barua, A., & Katyaini, S. (2017). Climate Change and Adaptation through the Lens of Capability Approach: A Case Study from Darjeeling, Eastern Himalaya. In I. Management Association (Ed.), Natural Resources Management: Concepts, Methodologies, Tools, and Applications (pp. 1351-1365). Hershey, PA: IGI Global. https://doi.org/ doi:10.4018/978-1-5225-0803-8.ch064

Mir, S. A., Shah, M. A., Mir, M. M., & Iqbal, U. (2017). New Horizons of Nanotechnology in Agriculture and Food Processing Industry. In B. Nayak, A. Nanda, & M. Bhat (Eds.), *Integrating Biologically-Inspired Nanotechnology into Medical Practice* (pp. 230–258). Hershey, PA: IGI Global. doi:10.4018/978-1-5225-0610-2.ch009

Moallem, M., Sterrett, W. L., Gordon, C. R., Sukhera, S. M., Mahmood, A., & Bashir, A. (2021). An Investigation of the Effects of Integrating Computing and Project- or Problem-Based Learning in the Context of Environmental Sciences: A Case of Pakistani STEM Teachers. In S. Schroth & J. Daniels (Eds.), *Building STEM Skills Through Environmental Education* (pp. 49–89). IGI Global. https://doi.org/10.4018/978-1-7998-2711-5.ch003

Mujere, N., & Moyce, W. (2017). Climate Change Impacts on Surface Water Quality. In W. Ganpat & W. Isaac (Eds.), *Environmental Sustainability and Climate Change Adaptation Strategies* (pp. 322–340). Hershey, PA: IGI Global. doi:10.4018/978-1-5225-1607-1.ch012

Mukherjee, A. B., Krishna, A. P., & Patel, N. (2018). Geospatial Technology for Urban Sciences. In *Geospatial Technologies in Urban System Development: Emerging Research and Opportunities* (pp. 99–120). Hershey, PA: IGI Global. doi:10.4018/978-1-5225-3683-3.ch005

Nagarajan, S. K., & Sangaiah, A. K. (2017). Vegetation Index: Ideas, Methods, Influences, and Trends. In N. Kumar, A. Sangaiah, M. Arun, & S. Anand (Eds.), *Advanced Image Processing Techniques and Applications* (pp. 347–386). Hershey, PA: IGI Global. doi:10.4018/978-1-5225-2053-5.ch016

Naraine, L., & Meehan, K. (2017). Strengthening Food Security with Sustainable Practices by Smallholder Farmers in Lesser Developed Economies. In W. Ganpat, R. Dyer, & W. Isaac (Eds.), *Agricultural Development and Food Security in Developing Nations* (pp. 57–81). Hershey, PA: IGI Global. doi:10.4018/978-1-5225-0942-4.ch003

Naraine, L., & Meehan, K. (2017). Strengthening Food Security with Sustainable Practices by Smallholder Farmers in Lesser Developed Economies. In W. Ganpat, R. Dyer, & W. Isaac (Eds.), *Agricultural Development and Food Security in Developing Nations* (pp. 57–81). Hershey, PA: IGI Global. doi:10.4018/978-1-5225-0942-4.ch003

Naraine, L., & Meehan, K. (2017). Strengthening Food Security with Sustainable Practices by Smallholder Farmers in Lesser Developed Economies. In W. Ganpat, R. Dyer, & W. Isaac (Eds.), *Agricultural Development and Food Security in Developing Nations* (pp. 57–81). Hershey, PA: IGI Global. doi:10.4018/978-1-5225-0942-4.ch003

Nikolaou, K., Tsakiridou, E., Anastasiadis, F., & Mattas, K. (2017). Exploring Alternative Distribution Channels of Agricultural Products. *International Journal of Food and Beverage Manufacturing and Business Models*, *2*(2), 36–66. doi:10.4018/IJFBMBM.2017070103

Nishat, K. J., & Rahman, M. S. (2018). Disaster, Vulnerability, and Violence Against Women: Global Findings and a Research Agenda for Bangladesh. In N. Mahtab, T. Haque, I. Khan, M. Islam, & I. Wahid (Eds.), *Handbook of Research on Women's Issues and Rights in the Developing World* (pp. 235–250). Hershey, PA: IGI Global. doi:10.4018/978-1-5225-3018-3.ch014

O'Hara, S., Jones, D., & Trobman, H. B. (2018). Building an Urban Food System Through UDC Food Hubs. In A. Burtin, J. Fleming, & P. Hampton-Garland (Eds.), *Changing Urban Landscapes Through Public Higher Education* (pp. 116–143). Hershey, PA: IGI Global. doi:10.4018/978-1-5225-3454-9.ch006

Obayelu, A. E. (2018). Integrating Environment, Food Systems, and Sustainability in Feeding the Growing Population in Developing Countries. In A. Obayelu (Ed.), *Food Systems Sustainability and Environmental Policies in Modern Economies* (pp. 1–14). Hershey, PA: IGI Global. doi:10.4018/978-1-5225-3631-4.ch001

Othman, R., Nath, N., & Laswad, F. (2018). Environmental Reporting and Accounting: Sustainability Hybridisation. In G. Azevedo, J. da Silva Oliveira, R. Marques, & A. Ferreira (Eds.), *Handbook of Research on Modernization and Accountability in Public Sector Management* (pp. 130–158). Hershey, PA: IGI Global. doi:10.4018/978-1-5225-3731-1.ch007

Ouadi, A., & Zitouni, A. (2021). Phasor Measurement Improvement Using Digital Filter in a Smart Grid. In A. Recioui & H. Bentarzi (Eds.), *Optimizing and Measuring Smart Grid Operation and Control* (pp. 100–117). IGI Global. https://doi.org/10.4018/978-1-7998-4027-5.ch005

Padigala, B. S. (2018). Traditional Water Management System for Climate Change Adaptation in Mountain Ecosystems. In I. Management Association (Ed.), Climate Change and Environmental Concerns: Breakthroughs in Research and Practice (pp. 630-655). Hershey, PA: IGI Global. doi:10.4018/978-1-5225-5487-5.ch033

Panda, C. K. (2018). Mobile Phone Usage in Agricultural Extension in India: The Current and Future Perspective. In F. Mtenzi, G. Oreku, D. Lupiana, & J. Yonazi (Eds.), *Mobile Technologies and Socio-Economic Development in Emerging Nations* (pp. 1–21). Hershey, PA: IGI Global. doi:10.4018/978-1-5225-4029-8.ch001

Pandian, S. L., Yarrakula, K., & Chaudhury, P. (2018). GIS-Based Decision Support System for Village Level: A Case Study in Andhra Pradesh. In S. Chaudhuri & N. Ray (Eds.), *GIS Applications in the Tourism and Hospitality Industry* (pp. 275–295). Hershey, PA: IGI Global. doi:10.4018/978-1-5225-5088-4.ch012

Quaranta, G., & Salvia, R. (2017). Social-Based Product Innovation and Governance in The Milk Sector: The Case of Carciocacio and Innonatura. In T. Tarnanidis, M. Vlachopoulou, & J. Papathanasiou (Eds.), *Driving Agribusiness With Technology Innovations* (pp. 293–310). Hershey, PA: IGI Global. doi:10.4018/978-1-5225-2107-5.ch015

Rahman, M. K., Schmidlin, T. W., Munro-Stasiuk, M. J., & Curtis, A. (2017). Geospatial Analysis of Land Loss, Land Cover Change, and Landuse Patterns of Kutubdia Island, Bangladesh. *International Journal of Applied Geospatial Research*, *8*(2), 45–60. doi:10.4018/IJAGR.2017040104

Rajack-Talley, T. A. (2017). Agriculture, Trade Liberalization and Poverty in the ACP Countries. In W. Ganpat, R. Dyer, & W. Isaac (Eds.), *Agricultural Development and Food Security in Developing Nations* (pp. 1–20). Hershey, PA: IGI Global. doi:10.4018/978-1-5225-0942-4.ch001

Rajamanickam, S. (2018). Exploring Landscapes in Regional Convergence: Environment and Sustainable Development in South Asia. In I. Management Association (Ed.), Sustainable Development: Concepts, Methodologies, Tools, and Applications (pp. 1051-1087). Hershey, PA: IGI Global. https://doi.org/doi:10.4018/978-1-5225-3817-2.ch047

Rizvi, S. M., & Dearden, A. (2018). KHETI: ICT Solution for Agriculture Extension and Its Replication in Open and Distance Learning. In U. Pandey & V. Indrakanti (Eds.), *Open and Distance Learning Initiatives for Sustainable Development* (pp. 163–174). Hershey, PA: IGI Global. doi:10.4018/978-1-5225-2621-6.ch008

Roşu, L., & Macarov, L. I. (2017). Management of Drought and Floods in the Dobrogea Region. In I. Management Association (Ed.), Agri-Food Supply Chain Management: Breakthroughs in Research and Practice (pp. 372-403). Hershey, PA: IGI Global. https://doi.org/ doi:10.4018/978-1-5225-1629-3.ch016

Rouzbehani, K., & Rouzbehani, S. (2018). Mapping Women's World: GIS and the Case of Breast Cancer in the US. *International Journal of Public Health Management and Ethics*, *3*(1), 14–25. doi:10.4018/IJPHME.2018010102

Roy, D. (2018). Success Factors of Adoption of Mobile Applications in Rural India: Effect of Service Characteristics on Conceptual Model. In M. Khosrow-Pour, D.B.A. (Ed.), Green Computing Strategies for Competitive Advantage and Business Sustainability (pp. 211-238). Hershey, PA: IGI Global. https://doi.org/ doi:10.4018/978-1-5225-5017-4.ch010

Sajeva, M., Lemon, M., & Sahota, P. S. (2017). Governance for Food Security: A Framework for Social Learning and Scenario Building. *International Journal of Food and Beverage Manufacturing and Business Models*, *2*(2), 67–84. doi:10.4018/ IJFBMBM.2017070104

Sambhanthan, A., & Potdar, V. (2017). A Study of the Parameters Impacting Sustainability in Information Technology Organizations. *International Journal of Knowledge-Based Organizations*, *7*(3), 27–39. doi:10.4018/IJKBO.2017070103

Sanga, C., Kalungwizi, V. J., & Msuya, C. P. (2017). Bridging Gender Gaps in Provision of Agricultural Extension Service Using ICT: Experiences from Sokoine University of Agriculture (SUA) Farmer Voice Radio (FVR) Project in Tanzania. In I. Management Association (Ed.), Discrimination and Diversity: Concepts, Methodologies, Tools, and Applications (pp. 682-697). Hershey, PA: IGI Global. https://doi.org/ doi:10.4018/978-1-5225-1933-1.ch031

Santra, A., & Mitra, D. (2017). Role of Remote Sensing in Potential Fishing Zone Forecast. In A. Santra & S. Mitra (Eds.), *Remote Sensing Techniques and GIS Applications in Earth and Environmental Studies* (pp. 243–257). Hershey, PA: IGI Global. doi:10.4018/978-1-5225-1814-3.ch012

Schaap, D. (2017). SeaDataNet: Towards a Pan-European Infrastructure for Marine and Ocean Data Management. In P. Diviacco, A. Leadbetter, & H. Glaves (Eds.), *Oceanographic and Marine Cross-Domain Data Management for Sustainable Development* (pp. 155–177). Hershey, PA: IGI Global. doi:10.4018/978-1-5225-0700-0.ch007

Seckin-Celik, T. (2017). Sustainability Reporting and Sustainability in the Turkish Business Context. In U. Akkucuk (Ed.), *Ethics and Sustainability in Global Supply Chain Management* (pp. 115–132). Hershey, PA: IGI Global. doi:10.4018/978-1-5225-2036-8.ch006

Segbefia, A. Y., Barnes, V. R., Akpalu, L. A., & Mensah, M. (2018). Environmental Location Assessment for Seaweed Cultivation in Ghana: A Spatial Multi-Criteria Approach. *International Journal of Applied Geospatial Research*, *9*(1), 51–64. doi:10.4018/IJAGR.2018010104

Sen, Y. (2018). How to Manage Sustainability: A Framework for Corporate Sustainability Tools. In I. Management Association (Ed.), Sustainable Development: Concepts, Methodologies, Tools, and Applications (pp. 568-589). Hershey, PA: IGI Global. doi:10.4018/978-1-5225-3817-2.ch026

Shamshiry, E., Abdulai, A. M., Mokhtar, M. B., & Komoo, I. (2015). Regional Landfill Site Selection with GIS and Analytical Hierarchy Process Techniques: A Case Study of Langkawi Island, Malaysia. In P. Thomas, M. Srihari, & S. Kaur (Eds.), *Handbook of Research on Cultural and Economic Impacts of the Information Society* (pp. 248–282). Hershey, PA: IGI Global. doi:10.4018/978-1-4666-8598-7.ch011

Sharma, Y. K., Mangla, S. K., Patil, P. P., & Uniyal, S. (2018). Analyzing Sustainable Food Supply Chain Management Challenges in India. In M. Ram & J. Davim (Eds.), *Soft Computing Techniques and Applications in Mechanical Engineering* (pp. 162–180). Hershey, PA: IGI Global. doi:10.4018/978-1-5225-3035-0.ch008

Silvestrelli, P. (2018). The Impact of Events: To Which Extent Are They Sustainable for Tourist Destinations? Some Evidences From Expo Milano 2015. In M. Risso & S. Testarmata (Eds.), *Value Sharing for Sustainable and Inclusive Development* (pp. 185–204). Hershey, PA: IGI Global. doi:10.4018/978-1-5225-3147-0.ch008

Silvius, G. (2018). Sustainability Evaluation of IT/IS Projects. In I. Management Association (Ed.), Sustainable Development: Concepts, Methodologies, Tools, and Applications (pp. 26-40). Hershey, PA: IGI Global. https://doi.org/ doi:10.4018/978-1-5225-3817-2.ch002

Singh, R., Srivastava, P., Singh, P., Upadhyay, S., & Raghubanshi, A. S. (2017). Human Overpopulation and Food Security: Challenges for the Agriculture Sustainability. In R. Singh, A. Singh, & V. Srivastava (Eds.), *Environmental Issues Surrounding Human Overpopulation* (pp. 12–39). Hershey, PA: IGI Global. doi:10.4018/978-1-5225-1683-5.ch002

Singh, R., Srivastava, P., Singh, P., Upadhyay, S., & Raghubanshi, A. S. (2017). Human Overpopulation and Food Security: Challenges for the Agriculture Sustainability. In R. Singh, A. Singh, & V. Srivastava (Eds.), *Environmental Issues Surrounding Human Overpopulation* (pp. 12–39). Hershey, PA: IGI Global. doi:10.4018/978-1-5225-1683-5.ch002

Srivastava, N. (2017). Climate Change Mitigation: Collective Efforts and Responsibly. In I. Management Association (Ed.), Natural Resources Management: Concepts, Methodologies, Tools, and Applications (pp. 64-76). Hershey, PA: IGI Global. https://doi.org/ doi:10.4018/978-1-5225-0803-8.ch003

Stanganelli, M., & Gerundo, C. (2017). Understanding the Role of Urban Morphology and Green Areas Configuration During Heat Waves. *International Journal of Agricultural and Environmental Information Systems*, *8*(2), 50–64. doi:10.4018/IJAEIS.2017040104

Stewart, M. K., Hagood, D., & Ching, C. C. (2017). Virtual Games and Real-World Communities: Environments that Constrain and Enable Physical Activity in Games for Health. *International Journal of Game-Based Learning*, *7*(1), 1–19. doi:10.4018/IJGBL.2017010101

Stone, R. J. (2017). Modelling the Frequency of Tropical Cyclones in the Lower Caribbean Region. In W. Ganpat & W. Isaac (Eds.), *Environmental Sustainability and Climate Change Adaptation Strategies* (pp. 341–349). Hershey, PA: IGI Global. doi:10.4018/978-1-5225-1607-1.ch013

Syed, A., & Jabeen, U. A. (2018). Climate Change Impact on Agriculture and Food Security. In A. Eneanya (Ed.), *Handbook of Research on Environmental Policies for Emergency Management and Public Safety* (pp. 223–237). Hershey, PA: IGI Global. doi:10.4018/978-1-5225-3194-4.ch012

Tam, G. C. (2017). The Global View of Sustainability. In *Managerial Strategies and Green Solutions for Project Sustainability* (pp. 1–24). Hershey, PA: IGI Global. doi:10.4018/978-1-5225-2371-0.ch001

Tam, G. C. (2017). Understanding Project Sustainability. In *Managerial Strategies and Green Solutions for Project Sustainability* (pp. 110–139). Hershey, PA: IGI Global. doi:10.4018/978-1-5225-2371-0.ch005

Tam, G. C. (2017). Perspectives on Sustainability. In *Managerial Strategies and Green Solutions for Project Sustainability* (pp. 53–76). Hershey, PA: IGI Global. doi:10.4018/978-1-5225-2371-0.ch003

Tang, M., & Karunanithi, A. T. (2018). Visual Logic Maps (vLms). In *Advanced Concept Maps in STEM Education: Emerging Research and Opportunities* (pp. 108–149). Hershey, PA: IGI Global. doi:10.4018/978-1-5225-2184-6.ch005

Taşçıoğlu, M., & Yener, D. (2018). The Value and Scope of GIS in Marketing and Tourism Management. In S. Chaudhuri & N. Ray (Eds.), *GIS Applications in the Tourism and Hospitality Industry* (pp. 189–211). Hershey, PA: IGI Global. doi:10.4018/978-1-5225-5088-4.ch009

Thanh Tung, B. (2021). Pharmacology and Therapeutic Applications of Resveratrol. In A. Hussain & S. Behl (Eds.), *Treating Endocrine and Metabolic Disorders With Herbal Medicines* (pp. 321–333). IGI Global. https://doi.org/10.4018/978-1-7998-4808-0.ch014

Tianming, G. (2018). Food Security and Rural Development on Emerging Markets of Northeast Asia: Cases of Chinese North and Russian Far East. In V. Erokhin (Ed.), *Establishing Food Security and Alternatives to International Trade in Emerging Economies* (pp. 155–176). Hershey, PA: IGI Global. doi:10.4018/978-1-5225-2733-6.ch008

Tiftikçigil, B. Y., Yaşgül, Y. S., & Güriş, B. (2017). Sustainability of Foreign Trade Deficit in Energy: The Case of Turkey. In N. Ray (Ed.), *Business Infrastructure for Sustainability in Developing Economies* (pp. 94–109). Hershey, PA: IGI Global. doi:10.4018/978-1-5225-2041-2.ch005

Tiwari, S., Vaish, B., & Singh, P. (2018). Population and Global Food Security: Issues Related to Climate Change. In I. Management Association (Ed.), Climate Change and Environmental Concerns: Breakthroughs in Research and Practice (pp. 41-64). Hershey, PA: IGI Global. https://doi.org/ doi:10.4018/978-1-5225-5487-5.ch003

Toujani, A., & Achour, H. (2018). A Data Mining Framework for Forest Fire Mapping. In I. Management Association (Ed.), Information Retrieval and Management: Concepts, Methodologies, Tools, and Applications (pp. 771-794). Hershey, PA: IGI Global. doi:10.4018/978-1-5225-5191-1.ch033

Tripathy, B., & K., S. B. (2017). Rough Fuzzy Set Theory and Neighbourhood Approximation Based Modelling for Spatial Epidemiology. In I. Management Association (Ed.), *Public Health and Welfare: Concepts, Methodologies, Tools, and Applications* (pp. 1257-1268). Hershey, PA: IGI Global. https://doi.org/ doi:10.4018/978-1-5225-1674-3.ch058

Trukhachev, A. (2017). New Approaches to Regional Branding through Green Production and Utilization of Existing Natural Advantages. In I. Management Association (Ed.), Advertising and Branding: Concepts, Methodologies, Tools, and Applications (pp. 1758-1778). Hershey, PA: IGI Global. doi:10.4018/978-1-5225-1793-1.ch081

Tsobanoglou, G. O., & Vlachopoulou, E. I. (2017). Social-Ecological Systems in Local Fisheries Communities. In G. Korres, E. Kourliouros, & M. Michailidis (Eds.), *Handbook of Research on Policies and Practices for Sustainable Economic Growth and Regional Development* (pp. 306–316). Hershey, PA: IGI Global. doi:10.4018/978-1-5225-2458-8.ch026

Tuydes-Yaman, H., & Karatas, P. (2018). Evaluation of Walkability and Pedestrian Level of Service. In I. Management Association (Ed.), Intelligent Transportation and Planning: Breakthroughs in Research and Practice (pp. 264-291). Hershey, PA: IGI Global. doi:10.4018/978-1-5225-5210-9.ch012

Uddin, S., Chakravorty, S., Ray, A., & Sherpa, K. S. (2018). Optimal Location of Sub-Station Using Q-GIS and Multi-Criteria Decision Making Approach. *International Journal of Decision Support System Technology, 10*(2), 65–79. doi:10.4018/IJDSST.2018040104

Uddin, S., Chakravorty, S., Sherpa, K. S., & Ray, A. (2018). Power Distribution System Planning Using Q-GIS. *International Journal of Energy Optimization and Engineering, 7*(2), 61–75. doi:10.4018/IJEOE.2018040103

Uzun, F. V. (2018). Natural Resources Management. In A. Eneanya (Ed.), *Handbook of Research on Environmental Policies for Emergency Management and Public Safety* (pp. 1–21). Hershey, PA: IGI Global. doi:10.4018/978-1-5225-3194-4.ch001

V, M., Agrawal, R., Sharma, V., & T.N., K. (2018). Supply Chain Social Sustainability and Manufacturing. In I. Management Association (Ed.), *Technology Adoption and Social Issues: Concepts, Methodologies, Tools, and Applications* (pp. 226-252). Hershey, PA: IGI Global. https://doi.org/ doi:10.4018/978-1-5225-5201-7.ch011

van der Vliet-Bakker, J. M. (2017). Environmentally Forced Migration and Human Rights. In C. Akrivopoulou (Ed.), *Defending Human Rights and Democracy in the Era of Globalization* (pp. 146–180). Hershey, PA: IGI Global. doi:10.4018/978-1-5225-0723-9.ch007

Vargas-Hernández, J. G. (2022). A Comprehensive Entrepreneurship Model for the Internationalization of Green Innovation Businesses. In U. Akkucuk (Ed.), *Disruptive Technologies and Eco-Innovation for Sustainable Development* (pp. 131–149). IGI Global. https://doi.org/10.4018/978-1-7998-8900-7.ch008

Vaskov, A. G., Lin, Z. Y., Tyagunov, M. G., Shestopalova, T. A., & Deryugina, G. V. (2018). Design of Renewable Sources GIS for ASEAN Countries. In V. Kharchenko & P. Vasant (Eds.), *Handbook of Research on Renewable Energy and Electric Resources for Sustainable Rural Development* (pp. 1–25). Hershey, PA: IGI Global. doi:10.4018/978-1-5225-3867-7.ch001

Vázquez, D. G., & Gil, M. T. (2017). Sustainability in Smart Cities: The Case of Vitoria-Gasteiz (Spain) – A Commitment to a New Urban Paradigm. In L. Carvalho (Ed.), *Handbook of Research on Entrepreneurial Development and Innovation Within Smart Cities* (pp. 248–268). Hershey, PA: IGI Global. doi:10.4018/978-1-5225-1978-2.ch012

Wahab, I. N., & Soonthodu, S. (2018). Geographical Information System in Eco-Tourism. In S. Chaudhuri & N. Ray (Eds.), *GIS Applications in the Tourism and Hospitality Industry* (pp. 61–75). Hershey, PA: IGI Global. doi:10.4018/978-1-5225-5088-4.ch003

Weiss-Randall, D. (2018). Cultivating Environmental Justice. In *Utilizing Innovative Technologies to Address the Public Health Impact of Climate Change: Emerging Research and Opportunities* (pp. 110–143). Hershey, PA: IGI Global. doi:10.4018/978-1-5225-3414-3.ch004

Weiss-Randall, D. (2018). Cultivating Resilience. In *Utilizing Innovative Technologies to Address the Public Health Impact of Climate Change: Emerging Research and Opportunities* (pp. 204–235). Hershey, PA: IGI Global. doi:10.4018/978-1-5225-3414-3.ch007

Weiss-Randall, D. (2018). Climate Change Solutions: Where Do We Go From Here? In *Utilizing Innovative Technologies to Address the Public Health Impact of Climate Change: Emerging Research and Opportunities* (pp. 236–268). Hershey, PA: IGI Global. doi:10.4018/978-1-5225-3414-3.ch008

Whyte, K. P., List, M., Stone, J. V., Grooms, D., Gasteyer, S., Thompson, P. B., . . . Bouri, H. (2018). Uberveillance, Standards, and Anticipation: A Case Study on Nanobiosensors in U.S. Cattle. In I. Management Association (Ed.), Biomedical Engineering: Concepts, Methodologies, Tools, and Applications (pp. 577-596). Hershey, PA: IGI Global. https://doi.org/ doi:10.4018/978-1-5225-3158-6.ch025

Wulff, E. (2017). Data and Operational Oceanography: A Review in Support of Responsible Fisheries and Aquaculture. In P. Diviacco, A. Leadbetter, & H. Glaves (Eds.), *Oceanographic and Marine Cross-Domain Data Management for Sustainable Development* (pp. 303–324). Hershey, PA: IGI Global. doi:10.4018/978-1-5225-0700-0.ch013

Yener, D. (2017). Geographic Information Systems and Its Applications in Marketing Literature. In S. Faiz & K. Mahmoudi (Eds.), *Handbook of Research on Geographic Information Systems Applications and Advancements* (pp. 158–172). Hershey, PA: IGI Global. doi:10.4018/978-1-5225-0937-0.ch006

Yu, K., Liu, Y., & Sharma, A. (2021). Analyze the Effectiveness of the Algorithm for Agricultural Product Delivery Vehicle Routing Problem Based on Mathematical Model. *International Journal of Agricultural and Environmental Information Systems, 12*(3), 26–38. https://doi.org/10.4018/IJAEIS.2021070103

Related References

Zhou, X., Sharma, A., & Mohindru, V. (2021). Research on Linear Programming Algorithm for Mathematical Model of Agricultural Machinery Allocation. *International Journal of Agricultural and Environmental Information Systems*, *12*(3), 1–12. https://doi.org/10.4018/IJAEIS.2021070101

About the Contributors

Olayinka Ohunakin is an internationally recognized Professor of Energy Systems. He is a Professor at the Department of Mechanical Engineering, Covenant University, Nigeria. He is a Beahrs Fellow, College of Natural Resources, University of California, Berkeley; Visiting Scholar, Center for African Studies, University of California, Berkeley; Visiting Professor, Centre for Petroleum, Energy Economics and Law (CPEEL), University of Ibadan, Nigeria; Visiting Professor, Faculty of Engineering & the Built Environment, University of Johannesburg, South Africa. He is also the Director and Co-founder, The Energy and Environment Research Group (TEERG), Department of Mechanical Engineering, Covenant University, Nigeria. Prof. Ohunakin is an energy and environmental expert with vast knowledge in renewable energy technologies, distributed generation, and hybrid grid systems, design and evaluation of low-carbon systems, energy modeling, climate modeling, energy economics, policies, finance, and management, with in-depth researches in energy conversion and management, energy efficiency and conservation techniques, conventional energy sources and power plants.

* * *

Donald Abonyi is a Masters Student.

Adejumoke Akinbusoye is a legal practitioner and a researcher with interests in the field of environment, development, energy and climate change law. She obtained her LL. B degree from the University of Lagos (2010) and her LL. M degree in Environmental Law and Policy from the Centre for Energy, Petroleum, Mineral Law and Policy, (CEPMLP) University of Dundee (2014). She was called to the Nigerian Bar in the year 2012. She is currently concluding her doctoral studies at the Department of Petroleum, Energy, Economics and Law, University of Ibadan. She teaches International Trade Law, Mineral and Water law and Petroleum and Energy Law at the University of Lagos, Nigeria. Her experience in legal practice spans litigation, corporate commercial practice, corporate governance, and legal advocacy. She is a member of the Nigerian Bar Association, The American Society of International Law, The Institute of Chartered Secretaries Administrators of Nigeria.

Alexander Akolo is a dedicated professional with demonstrated experience in regulatory compliance, government relations, contracts administration and settlements, negotiations, research and reportage in the Nigerian Electricity Supply Industry. He has recently joined Nextier Power (a subsidiary of Nextier Capital Limited) as a Senior Energy Consultant. Before this role, Alexander was a Regulatory Affairs and Government Relations Manager at Nigeria's largest electricity distribution company, Ibadan Electricity Distribution Company Plc. Alexander has deep knowledge of the plethora of regulations and directives in the NESI. He has worked as a Technical Assistant at the Bureau of Trade and Investment and as a Contracts Management Analyst at the Embassy of Brazil and the Nigerian Bulk Electricity Trading Plc, respectively. Alexander is about to complete a Doctor of Philosophy (Ph.D.) in Energy Studies (Renewable Energy Finance) from the University of Ibadan. He holds a M.Sc. Energy Studies degree from the University of Ibadan and a B.Sc. Economics degree from the University of Jos. He has attended several energy conferences within and outside Nigeria. He is a member of the Nigerian Association of Energy Economics (NAEE), the Global Association of Risk Professionals (GARP), the Society of Petroleum Engineers (SPE), and the Energy Institute.

Mark Akrofi holds a Ph.D. in Sustainability Science from the United Nations University Institute for the Advanced Study of Sustainability, Tokyo, Japan. His research explores the governance, policies, and processes of sustainable energy transitions, with a regional focus on Africa.

Niyonzima Damascene is a Renewable Energy Expert.

Axel Nguedia-Nguedoung holds a MSc. in Energy Engineering from the Pan African University Institute of Water and Energy Sciences, in Algeria. His MSc. thesis investigated the use of Open Source tool (OSEMOSys) in the optimization of energy systems and investment planning in Africa, in the context of climate change. Prior to completing his MSc, Axel worked as a research intern at the African Climate Policy Centre, analysing the energy contributions of Central African Countries' Intended Determined Contributions (INDCs) in the frame of the Paris Agreement. Axel worked in the energy consultancy industry in Cameroon, as an independent energy consultant, where he contributed to increasing energy access in his community by offering off-grid energy solutions and energy efficiency services. These services included energy audits of institutional buildings, design and sizing of solar home systems, as well as design and economic analysis of biomass wastes to biogas. He also contributed to the development of the International Union for Conservation of the Nature's (IUCN) priorities in the energy sector in Cameroon. Axel research's interest is in energy access with specific focus on sustainable use of renewable energy resources. He is further interested in the use of Information and Communication Technologies in the fields of renewable energies and energy efficiency.

Sanderine Nonhebel is an Associate Professor, Faculty of Science and Engineering, Integrated Research on Energy, Environment and Society (IREES), University of Groningen.

Nima Norouzi does research in techno-economic, techno-environmental, techno-social, and legal aspects of Energy and natural resources economics.

Dilinna Lucy Nwobi is a McArthur foundation Scholar at the Centre for Petroleum Energy Economics and Law (CPEEL) (Now Department of Minerals, Petroleum, Economics and Law) at the University of Ibadan. A multidisciplinary researcher with a bachelor's degree in Computer science, master's in Computer systems and master's in Energy Finance respectively. She has been part of several research works including The Ajibode Household Energy Consumption Survey and the Household Energy and Health Workshop where she headed the survey (instrument) design and administration team. Her PhD research area is Energy finance with a particular interest in the dynamics of Public Finance and the Macroeconomics of Government Budgets. Her research objective is to evaluate the existing variance, relationship and causality between the Federal fiscal budget and the actuals of Nigeria's budget and make forecasts based on historic data from the FGN Budget framework. The contribution of this research to knowledge is an insight into the dynamics of Nigeria's government budget design and performance based on revenue sources and the budget forecast options using empirical methods and data. Her other interests include the Energy Policy formulation to ensure sustainability via energy security in Sub-Saharan Africa. She is currently a member of various professional bodies, including the Nigerian Association of Energy Economists and the Nigerian Computer Society.

Olutosin Ogunleye is a Senior Research Fellow at the Centre for Innovation and Creativity, Nigerian Defence Academy. He obtained B.Eng in Electrical/Electronic Engineering and M.Eng Electronic/Communications Engineering from the Nigerian Defence Academy in 1998 and 2012 respectively. He also bagged a doctorate in Energy Studies from the Faculty of Multidisciplinary Studies, University of Ibadan, Nigeria in 2021. He has attended energy conferences within and outside Nigeria and his research interests include solar thermal systems and energy security. He is a corporate member of the Nigerian Society of Engineers, Nigeria Institute of Management and a chartered engineer of the Council for the Regulation of Engineering in Nigeria. He is also a fellow of the Army War College (fwc) Nigeria and fellow of the National Defence College (fdc) Nigeria.

Olayinka S. Ohunakin is an internationally recognized professor of energy systems. He is a Professor at the Department of Mechanical Engineering, Covenant University, Nigeria. He is a Beahrs Fellow, College of Natural Resources, University of California, Berkeley; Visiting Scholar, Center for African Studies, University of California, Berkeley; Visiting Professor, Centre for Petroleum, Energy Economics and Law (CPEEL), University of Ibadan, Nigeria; Senior Research Associate, Faculty of Engineering & the Built Environment, University of Johannesburg, South Africa. He is also the Head, The Energy and Environment Research Group (TEERG), Department of Mechanical Engineering, Covenant University, Nigeria. He is an energy and environmental expert with vast knowledge in renewable energy technologies, distributed generation and hybrid grid systems, design and evaluation of low-carbon systems, energy modeling, climate modeling, energy audits, cost-benefit analysis of renewable energy investments, energy efficient building simulations and designs, energy economics, policies, finance and management, with in-depth researches in energy conversion and management, energy efficiency and conservation techniques, conventional energy sources and power plants. He is a registered member of several professional bodies; he has written several articles widely published on energy in leading peer-reviewed 'high impact' journals cutting across Springer, Elsevier, Sage and Taylor & Francis publishers (). He is currently serving in the capacity as the Conference Chair, International Conference on Engineering for a Sustainable World (ICESW 2021), and reviewer/editorial membership to numerous energy journals. He has also presented several positions papers in conferences and workshops both locally and internationally. He has been involved in several National Assignments for Nigeria at the United Nations Framework Convention for Climate Change (UNFCCC), member, Inter-Ministerial Committee on Climate Change (ICCC), and technical expert on standards for the Standard Organization of Nigeria. He has served as a consultant on several projects for the Worldbank/IFC, AfDB, KfW, AFD, GIZ, Japan International Cooperation Agency (JICA), UNDP-GEF project on Energy Efficiency in Nigeria, etc. He is also a consultant to several NGOs and private bodies (both foreign and local). He is currently a Nominated National Expert for Nigeria at UNFCCC.

Oluwatomiwa Phillips is an avid researcher who has degrees in geography and urban management and development as well as interests in the social side of urban waste management, energy and climate change She obtained her MSc in urban management and development from the Erasmus University in the Netherlands (2014) and her BSc in geography from the University of Ibadan (2010). She is currently working on her PhD with a focus on household cooking energy at the department for minerals, petroleum, energy economics and law in the University of Ibadan.

Ilhem Rabehi is a PhD Student.

Ahmad Tasnim Siddiqui is an Associate Professor at the Department of Computer Science & Applications, Babu Banarasi Das University, Lucknow, India. He is Ph.D. in Computer Science. He also holds a Master of Computer Applications, and an M. Phil (Computer Science) from Madurai Kamaraj University, Madurai, India. He has many publications in reputed journals indexed at SCI, SCIE, and SCOPUS. He has also published a book chapter in EMERALD insight and one chapter is about to release by SPRINGER. His research interest includes e-commerce, e-learning, active learning, web mining, IoT, ICT, e-health, and cloud computing. He has a total of 15+ years of experience including 4.7 years of software industry experience. His favorite subjects are E-commerce and web Technologies using .net. His research work can be seen on Google Scholar profile

João Simões is an Assistant Professor at City University of Macau, China. He earned his Ph.D. in Portuguese-speaking Countries Studies from the same university, a Master's degree in Chinese Studies from the University of Aveiro, Portugal, and a Bachelor's degree in Electrical and Computer Engineering from the University of Lisbon, Portugal. He co-edited the book 'Changing the Paradigm of Energy Geopolitics: Security, Resources, and Pathways in Light of Global Challenges,' published by Peter Lang Publishers in 2023.

Nupur Soni is Associate Professor in the School of Computer Applications, Babu Banarasi Das University, Lucknow, India. She has extensive experience in academics. Her interest of research is Cloud computing, IoT, Security, Blockchain etc. She has published many papers in scopus indexed International Journals as well as IEEE conferences.

Djiby Thiam is an Associate Professor, School of Economics, University of Cape Town.

Thomas van Huyssteen is a PhD candidate at the University of Cape Town and the University of Groningen.

Index

B

C

D

E

R

S

www.ingramcontent.com/pod-product-compliance
Lightning Source LLC
LaVergne TN
LVHW080645011026
849160LV00014B/1326